ELECTRONICS AND WIRING
FOR MODEL RAILWAYS

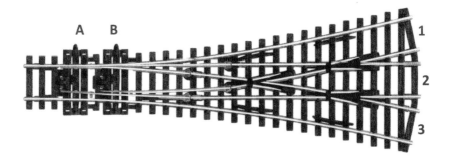

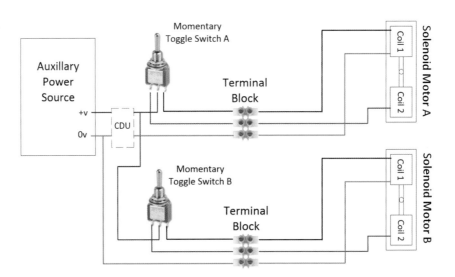

ELECTRONICS AND WIRING
FOR MODEL RAILWAYS

ANDREW DUCKWORTH

THE CROWOOD PRESS

First published in 2019 by
The Crowood Press Ltd
Ramsbury, Marlborough
Wiltshire SN8 2HR

enquiries@crowood.com

www.crowood.com

© Andrew Duckworth 2019

All rights reserved. No part of this publication may be reproduced or transmitted in any form or by any means, electronic or mechanical, including photocopy, recording, or any information storage and retrieval system, without permission in writing from the publishers.

British Library Cataloguing-in-Publication Data
A catalogue record for this book is available from the British Library.

ISBN 978 1 78500 623 4

Typeset by Servis Filmsetting Ltd, Stockport, Cheshire
Printed and bound in India by Parksons Graphics

CONTENTS

INTRODUCTION		6
CHAPTER 1:	**THE LAYOUT**	12
CHAPTER 2:	**ELECTRONIC AND ELECTRICAL COMPONENTS EXPLAINED**	29
CHAPTER 3:	**ELECTRICAL SYSTEMS AND POWER SUPPLIES**	45
CHAPTER 4:	**WIRE AND CABLES**	52
CHAPTER 5:	**CONTROL PANELS**	60
CHAPTER 6:	**SWITCHES**	66
CHAPTER 7:	**TRACK WIRING**	80
CHAPTER 8:	**SIGNALS**	106
CHAPTER 9:	**LIGHTING PROJECTS**	115
CHAPTER 10:	**DETECTION**	123
CHAPTER 11:	**TESTING AND TROUBLESHOOTING**	135
APPENDIX I:	**GLOSSARY OF TERMS**	139
APPENDIX II:	**ELECTRONIC SYMBOLS**	144
APPENDIX III:	**MODEL RAILWAY TRACK SYMBOLS**	152
APPENDIX IV:	**CAPACITOR CONVERSION CHART**	154
APPENDIX V:	**ENAMELLED COPPER WIRE**	156
INDEX		158

INTRODUCTION

My love of model railways started way back in 1962 with my grandmother sending me a Hornby OO kit for my twelfth birthday. I lived in Nyasaland (now Malawi) in those days, and this was the time when post used to take about a month to arrive, and parcels took about two months. There were no mobiles – you had to book a telephone call twenty-four hours in advance and it cost a week's pocket money, or send a telex that was converted in the UK to a telegram, again costing a fortune: so ordering was very difficult. So I resorted to sending requests by post to my grandmother, who would send me parts for my layout three times a year, on my birthday, for the summer holidays, and at Christmas. Over the next four years my first railway grew from a small kit into a large four- by three-metre U-shaped layout. This was, of course, the era when the track was a three-rail system, and all parts were made in metal.

During this period I developed a desire for anything electrical and electronic – although the transistor had only just been invented and there were no integrated circuits, it all fascinated me. After my school years I went to a polytechnic in Johannesburg to do a course in electrical/electronic engineering. On completion of the course and with certificate in hand, I applied for jobs in the UK. In my naïvety I stated that I had a love for model railways, thinking this would help with my job application – little did I know. My first job application I received was from a company in Plymouth called ML Engineering. I flew to the UK to start my working career at the age of twenty-one.

ML Engineering was an American company based in Plymouth, which designed and manufactured electrical systems to electrify British Rail. They designed and manufactured an 'automatic train detection system' (AWS), which looked like a big black rectangular box placed between the rails. This sent a signal to the train and back to the signal box, indicating that a train had just passed over it, as shown in Fig. Int.1. The company also manufactured the relay interlocking system between the train detector, the points and the signals – Fig. Int.2.

The relay room for most stations was a two-storey building with the relay room under the signal box, and it had racks of relays. Each relay was interlocked with others, so when a points button was pressed on the mimic panel, dozens of relays had to be in the correct position before the current would

Fig. Int.1   Train sensor.

Fig. Int.2   Relay room. COMPETENCY AUSTRALIA

INTRODUCTION

Fig. Int.3   Mimic display signal box.

pass through them to the point motor. This would also put all the required signals in the correct aspect. This was then put on to a mimic display panel, which replaced the levers in the signal box. In my time with the company I worked on Shenfield, Grantham and Crewe signal systems, converting them to electrical mimic displays with push buttons to control the points and the signals, as shown in Fig. Int.3

I remained with ML Engineering until 1976, when I returned to Nyasaland, now Malawi, to set up my own electronic manufacturing company. I was unable to use my knowledge in the new company, so branched out into other electronic products that would sell in that country, such as AM/FM radios, amplifiers and speed detectors. This was when my original model railway system was brought out of mothballs and rebuilt in my new home. By this time things had changed a great deal in the model railway business. It was now time to update my layout to the two-rail system, again a slow process as everything had to be ordered from the UK – but the mail system had improved to seven days, so not so bad.

In 1996 I came back to the UK with model railway system in tow. By this time there were transistors, integrated circuits and microprocessors – some of you may remember the Sinclair ZX80 and ZX81 computers introduced in 1981; also in the eighties came Amstrad computers CPC664 and CPC6128. Although very simple by modern standards, they could be easily programmed to do certain functions, a little like the Arduino and Raspberry Pi, to name a few, of today.

Back to the important bit.

MODEL RAIL GAUGE

There is a difference between gauge and scale. The gauge is the distance between the rails, from the inside of one rail to the inside of the other rail, with trains and rolling stock built for each particular gauge. Scale is the proportion that the size of the

INTRODUCTION

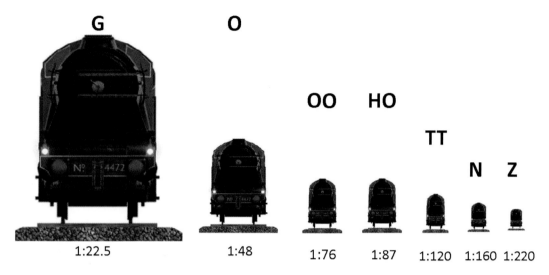

Fig. Int.4 Train size comparison.

model is compared to its real-world equivalent. It is normally expressed as a ratio (1:16 or 1/16) or a size (1in:1ft).

When a model train is scaled down the gauge is not necessarily to scale, but to the nearest standard gauge. This means that you could have

Train size comparison chart

Name	Scale	Description
Z	1:220 scale	Based on Marklin factory standards
N	2mm = 1ft 1:148 scale 9mm gauge	Twice as small as OO gauge
TT	3mm = 1ft 1:101.6 12mm gauge	This gauge originated in the USA, and was also produced at 2.5mm to 1ft, 1:120 scale. Enthusiasts using this scale need specialist support through the Three Millimetre Society
HO	3.5mm = 1ft 1:87 16.5mm gauge	This is the major gauge used outside the UK. At 3.5mm to 1ft, the track gauge at 16.5mm is virtually exact to scale for the standard gauge. When using this gauge it must not be confused with OO gauge: HO gauge is almost 15 per cent smaller. One can run HO gauge rolling stock on OO gauge layouts, the track gauges both being 16.5mm, but the difference in scale will immediately become very obvious
OO	4mm = 1ft 1:76 16.5mm gauge	This is the most popular scale for British modellers, and is probably the best supported in the industry with a wide range of ready-to-run models, kits and accessories
O	7mm = 1ft 1:43.5 32mm gauge	This scale has become more popular due to the availability of a large range of quality locomotive and rolling-stock kits. Technically the inside track width of 32mm is 3 per cent under scale
G	13.55 = 1ft 1:22.5 45mm gauge	G is generally used for garden railways of narrow gauge prototypes, and uses the same track gauge as 1 gauge. The scale ranges approximately from $1/19$ to $1/29$, according to the size and gauge of the prototype.

How to resize scale models using either a copier or printer

	G Scale	O Scale	S Scale	OO Scale	HO Scale	TT Scale	N Scale	Z Scale
G Scale		213%	284%	339%	386%	533%	711%	977%
O Scale	47%		133%	158%	181%	250%	333%	458%
S Scale	35%	75%		119%	136%	188%	250%	344%
OO Scale	30%	63%	84%		115%	158%	211%	289%
HO Scale	26%	55%	73%	87%		138%	184%	253%
TT Scale	19%	40%	53%	63%	73%		133%	183%
N Scale	14%	30%	40%	48%	54%	75%		138%
Z Scale	10%	22%	29%	35%	40%	55%	73%	

two different trains both with the same gauge, but in a slightly different scale. In practice this will be hardly noticeable, but it is worth bearing in mind.

The above table shows how to resize scale models using either a copier or printer: use the table to give the appropriate enlargement or reduction to rescale the drawing.

Find your scale in the table along the top, then scroll down to the desired scale and find out the factor you need to enlarge or reduce. So if, say, I have HO-scale plans that I want to enlarge to O-scale, I run across the top to HO, then down to O scale, and see that I need to enlarge the plans to 181 per cent. If I have O-scale plans that I want to reduce to S-scale, I run across the top to O and down to S, and see I need to reduce the plans to 75 per cent.

DC OR DCC

DC = DIRECT CURRENT

The instructions sent from the 'controller' are to vary the voltage to the track: this either speeds up the locomotive or slows it down. In the old days this was called a 'reostat', which simply changed the voltage from 0v to 12v, which in turn caused the locomotive engine to speed up or slow down. Modern speed controllers are called 'pulse-width modulation' regulators (PWM). PWM is a modulation technique used to convert a voltage into a pulsing signal. This modulation allows the control of the power supplied to electrical devices, especially to loads such as motors. The pulse-width modulation speed control works by driving the motor with a series of 'on-off' pulses and by varying the duty cycle – the fraction of time that the output voltage is 'on' compared to when it is 'off' – of the pulses while keeping the frequency constant. This will be explained in detail in the Power Supplies section.

DCC = DIGITAL COMMAND CONTROL

The instructions are sent from the 'controller' to the decoder on the locomotive, or any other item that has a decoder, by means of digital signals that are superimposed over a constant track voltage. In a DCC system the rails have a constant voltage running through them, as does all ancillary equipment. In this voltage there is a series of pulses that are sent to all the decoders on the layout. Each decoder is programmed to respond to a certain sequence of pulses: when it receives that sequence it actions the equipment. Put simply, you control the locomotive and not the track.

THE DC SYSTEM

The DC system has been in existence since the 1950s, and is probably the most common system used, due to the fact that it has been around for such a long time.

You cannot successfully run two trains on the same track. I say 'successfully' because you can run two trains on one track, but the problem is that, depending on the motors and load, they may well catch each other up, because they are both getting the same amount of voltage. There is no way to control them separately. Each track circuit will require its own speed controller, and each junction between tracks will require isolators, otherwise all trains will move in the same direction when power is applied to the track. This is why most points, even today, isolate the straight from the turn out – hence the reason you need to fit shorting clips to existing points to run DCC.

You cannot run interior carriage lights on a DC system as they will go out when the train is stationary, or will dim as the train slows down: you would have to fit a little battery pack inside the carriage. (*See* Lighting Projects.)

The points, turntables and any ancillary equipment requires a separate power source of 12–16v DC in most cases, and a bank of switches is needed to control these items.

This being said, you can run many trains on a DC system as long as each set of tracks has a separate speed controller. So you could have four circular tracks with a train running on each, with each going at different speeds and directions, and also work a shunting (fiddle) yard at the same time, as long as each is isolated from the other by points of isolating track. This is where your bank of switches comes in, to control the points and isolating track.

THE DCC SYSTEM

The DCC system was first introduced in the 1970s, but it was not very good so did not last long. The next generation has been around since the 1990s. In a DCC system the complete layout is

> **PROS AND CONS OF THE DC SYSTEM**
>
> **Pros:**
> - It's been around a long time and is tried and tested
> - It requires research if you want to do something out of the ordinary
> - It costs less than a DCC
> - It is very 'hands on'
> - It is more like the 'real thing' if you are 'in' to steam locos
>
> **Cons:**
> - There is a mass of wiring under the baseboard
> - You have a bank of switches that need to be labelled otherwise there will be chaos
> - Each track is almost like a separate system within your layout

live at 12v DC, and this voltage does not vary; what does vary is the signal, which is 'piggy-backed' on the voltage.

Each locomotive, set of points, and every controllable item has a 'decoder'. Each decoder has a specific address, which you create with the control station. So now the controller knows all the addresses of each locomotive, set of points, and so on. The decoder on the locomotive reads the instruction, checks it is for that loco, and acts on the instruction. This means that you could have more than one train on the same line: you could have one train in the station and another approaching the station, or you could reverse one train on the same line as a train is going forwards.

The fiddle yard can be worked with more than one train at the same time, as long as you have eyes in the back of your head.

The DCC system does away with the banks of switches, and electronic circuits to control this and that. The complete layout can be run from a computer or android with a display of your layout in real time.

PROS AND CONS OF THE DCC SYSTEM

Pros:
- A lot less wiring, requires a bus bar and spurs to each piece of equipment (points and so on)
- More control of locomotives and ancillary equipment
- Computer control, therefore you can create complete scenarios of your layout. Most old locomotives can be fitted with a decoder

Cons:
- The major downside is the cost. The average cost of a controller is around £200, though prices are on the way down
- Each locomotive, set of points and so on will cost around £15 for each decoder
- The decoder cost is on top of your normal layout costs

This is a very brief description of the two systems; however, there are thousands of pages on the web about both types.

For what it's worth I am a traditionalist and therefore would go for the DC system. Part of the magic of a model rail system is the fiddling, which never ends with a DC system – you do, however, need a greater knowledge of electricity and electronics.

CHAPTER ONE

THE LAYOUT

The layout is a very personal thing, whether it be a simple oval with sidings, or a replica of a real-life railway layout. The next big decision is to go either DC or DCC. A DC system is more commonly referred to as an 'analogue control system', where the two rails are powered by a 12-volt DC speed controller, and one rail is the +12v (feed) and the other rail is the 0v (return).

This book concentrates on the DC system; if you are going down the DCC route there are plenty of books on this subject. In a DC system the controller is fed to the rails so that we have one rail as the positive rail (feed), which we will show in this book as red, and one negative rail (return), which we will show as blue. In wiring an analogue layout there are three distinct sections: the track, the points, and all ancillary equipment.

SCALE

The next decision is to decide which scale you are going to work with. In this chapter we show all the model railway scales that are available. Certain manufacturers have track parts that are unique to them, and in some cases they can be used with other manufacturers of the same scale, but care must be taken with the angles of the curves and turnouts. Your first decision is where your layout is going to be housed, and how much room you have available. We have set out below a minimum curve for the most popular scales.

The reason there are two radii in certain scales is due to certain manufacturers having different radii for their layouts. When planning your layout it is important you decide which system you are opting for – you can then decide on the space required. These are the minimum, so if you want more than one curve – for example an inner and outer line – you will need more front-to-back space.

Gauge	Minimum radius
G scale	609–1,117mm
On3 scale	914mm
O scale	437–607mm
S scale	508mm
HOn3 scale	381mm
HO scale	381–558mm
OO scale	371mm
N scale	247mm
Z scale	195mm

There are plastic stencils available for most scales, which makes it very easy to design your layout on paper first. This is well worth doing, as it should show up any problems you may come across.

It is important to know the track spacing for your scale: if this is not maintained, trains can sideswipe each other at points or on curves.

TRACK SEPARATION

Each scale will have a different parallel track separation. The spacing can be made greater by putting straight rails between the two sets of points. In

Fig. 1.1 Adjustable track separator.

THE LAYOUT

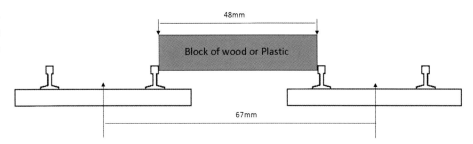

Fig. 1.2 Home-made track separator.

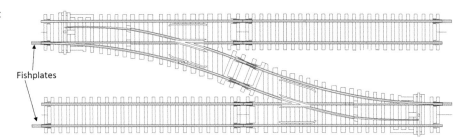

Fig. 1.3 Track separation.

some cases the spacing may have to be greater on the curves due to the overhang of certain coaches and locomotives. You will need to keep an eye on this, as there are no hard and fast rules. There are templates available for most gauges, or you could make one yourself.

TRACK JOINING

Most track sections come with a 'fishplate' on one rail at each end. These are used to lock each section physically and electrically together. You must ensure that the fishplate has been fitted correctly otherwise one rail will be higher than the other.

TRACK LENGTH

It is important to remember that track length as quoted by the manufacturer/supplier is the length of the track, and does not include the fishplate joiner.

MEASURING TRACK LENGTH

While you are designing your layout on paper ignore the fishplates; the length of any piece is one end of the rail to the other end of the rail, and this will give a true dimension of your layout.

WHY IS IT CALLED A FISHPLATE?

A fishplate is a metal bar that is bolted to the ends of two rails to join them together in a track. The name is derived from *fish*, a wooden bar with a curved profile used to strengthen a ship's mast. The top and bottom edges are tapered inwards so the device wedges itself between the top and bottom of the rail (which is called the 'fishing'), when it is bolted into place. In model railways, a fishplate is often a small copper or nickel silver plate that slips on to both rails to provide the functions of maintaining alignment and electrical continuity.

Fig. 1.4 Fishplates. HORNBY

14 THE LAYOUT

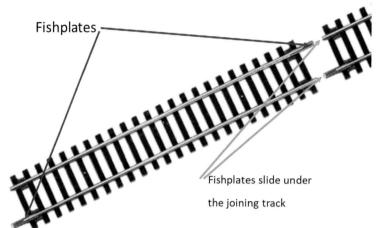

Fig. 1.5 Joining two tracks.

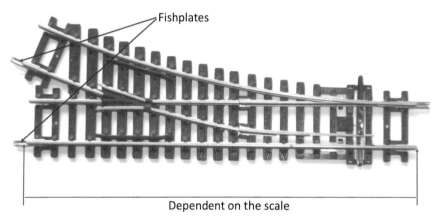

Fig. 1.6 Track length.

TRAIN AND COACH OVERHANG

As locomotives and coaches go around bends they create overhang and underhang. This means that placing structures and objects too close to the track will cause problems. Please bear in mind this is an average, and certain locos and carriages may have a greater overhang than the average.

Fig. 1.7 shows the overhang and underhang (red) for each of four curve radii. The red lines above and below the black lines (tracks) show the extent of the overhang or underhang. The table shows the distance in mm.

Train overhang and underhang

Curve	Underhang	Overhang
4th radius curve 572mm	8.00mm	7.50mm
3rd radius curve 505mm	8.50mm	8.00mm
2nd radius curve 438mm	9.50mm	9.00mm
1st radius curve 371mm	11.00mm	10.5mm

It must be stressed that these figures are for OO gauge, and will vary depending on the gauge you select.

THE LAYOUT

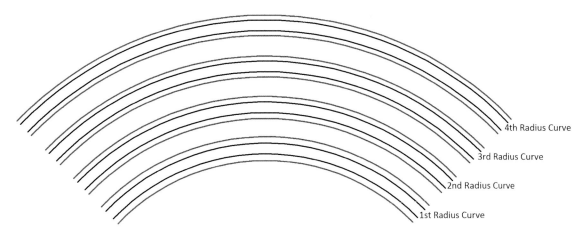

Fig. 1.7 *Overhang and underhang.*

TRACK RAMPS

If you are thinking of having dual-level tracks then you will need to create up and down ramps. The main problem here is the ability of the train to go up the ramp, and with how many coaches; the weight of the train is all important, as it needs to create friction with the track – the lighter it is, the more chance you will get wheel spin and go nowhere. A steep rise to 80mm will take up 1,344mm, but should work for most modern-made locomotives; however, a better ramp length would be 1,680mm. Using a Helices ramp will take up less space but will require a slower ascent and descent. Keeping track gradients at 2 per cent or below is a good rule of thumb. It can also look more realistic than a really steep gradient (as long as you have the space). Most manufacturers have their own ramp systems.

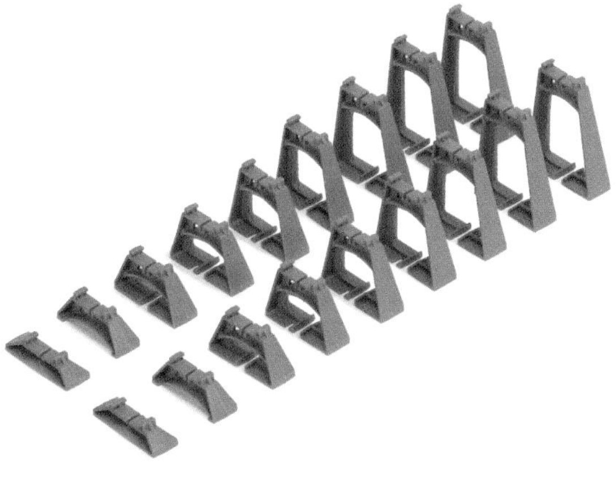

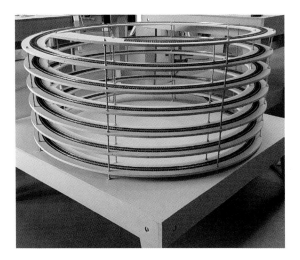

Fig. 1.8b *Helices.* MODEL RAILWAY BASEBOARDS

BANKING AND TILT

On a model railway layout, super-elevation serves no functional purpose: the weights and forces just don't work the same way as on the real thing. It's purely there for visual effect. Elevation (E) between rails will be slightly less than shim height if the shim

Fig. 1.8 *Set of plastic ramps.* HORNBY

Fig. 1.9 Virgin train banking at speed.

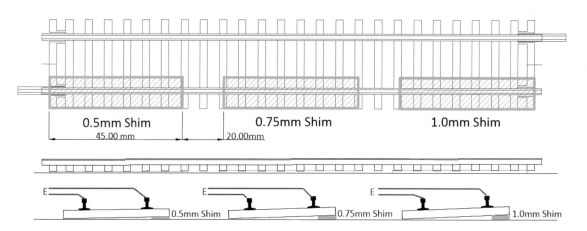

Fig. 1.10 Shims under the track.

is directly below the rail, but consistency of placement is more important. Shims should be increased by 0.25mm up to a maximum of 2.5mm – anything above this could cause problems with stationary trains on the curve. Shims should not be used on first and second radius curves.

If you want to go down this path you can get a pack of spacers at 0.25mm increments from your local builders' merchant. They just need to be cut in half to fit under the track.

Irrespective of the scale you decide upon, the basic principle of a DC system will be the same. In this chapter we are going to look at the polarity of the rails and how they change at points, or crossovers, and how they can come back on themselves if you are not careful. Let's start with the basic layout, an oval, as this is the simplest and easiest when starting to understand the principles of electrical wiring.

SINGLE RECTANGULAR LAYOUT

The layout will be connected to a speed controller, which will feed positive power to the outer rail and negative (0v) power to the inner rail. The locomotive

THE LAYOUT | 17

Fig. 1.11 Basic rectangular layout.

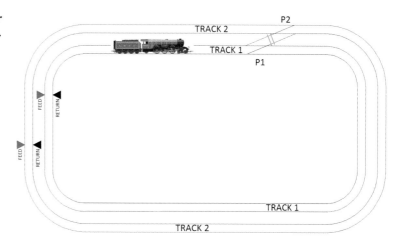

Fig. 1.12 Dual rectangular layout.

will now travel clockwise around the track. If the polarity is reversed, so that *feed* is the inner track and *return* is the outer track, the locomotive will travel anti-clockwise.

This is not very interesting as a layout, so we must now introduce some more tracks. From here on we will assume that the normal direction of travel is clockwise, as this will determine how we place our first set of points. We are now going to introduce a second oval outside the existing oval. To do this we must place a set of points in the track to allow the locomotive to move from track 1 to track 2, shown in Fig. 1.12 as 'P1' and 'P2'.

In Fig. 1.12 we have introduced a second oval on the outside of track 1, and this means we need to put at least two sets of points in place: P1 and P2. This allows us to get from track 1 to track 2, and vice versa. You will notice the two orange lines half way between the points: this is where the points will be isolated. When the points are straight through there is no electrical connection between track 1 and track 2, so to run a train on both tracks you would need two speed controllers, or one speed controller connected to both tracks.

POINTS

A set of points – also called a railroad switch or turnout – is used to guide a train from one track to another, such as a spur or siding. Basic points are either left hand or right hand. Other points are high speed points, normally twice as long as the standard point set, and curved points. Non left or right points are also available, and come as double slip or

18 THE LAYOUT

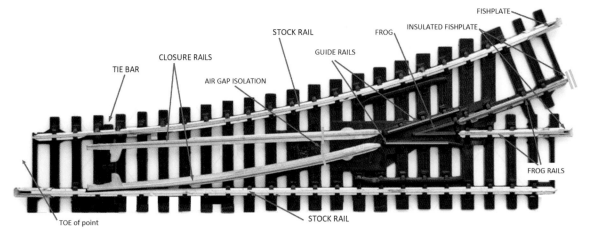

Fig. 1.13 Parts of a points set.

triple slip, giving two or three outlets from the single incoming direction.

Points are also divided into insulated frogs (the frog is insulated from the power) or electrofrogs (the frog is connected to the 'closure rails').

We will now look at a set of points to see how it works electrically.

In Fig. 1.13 we can see the different elements of a set of points. The 'tie-bar' is what the motor connects to; 'stock rails' are the outer non-moving rails; 'closure rails' are the moving rails that change the direction of travel and also change the path of electricity.

INSULATED FROG

The term 'frog' is taken from the part of a horse's hoof that it most closely resembles. The frog, also known as the common crossing, is the crossing point of two rails. Frogs can be live (made from metal) or dead (made from plastic), and note that not all metal frogs are live. The reason for all the talk about frogs is that the frog is an electrically dead area that can cause problems with short locomotives, because as the wheels pass over the frog area there is no power going to the locomotive engine, so the train can stop or judder.

Fig. 1.14 Detail of the points frog.

THE LAYOUT 19

Fig. 1.15 Detail of the manufacturer's connection at the frog.

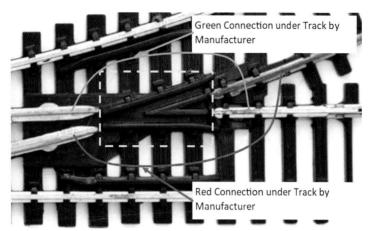

SELF-POWERED

With the self-powered type, the frog is electrically connected to the closure rails, so the position of this rail decides the polarity of the frog. Looking at Fig. 1.16, P1 – the closure rail – is connected to the red rail so the frog would be red. In Fig. 1.17, P1 the closure rail is connected to the blue rail so the frog would be blue. It is important to ensure that insulated fishplates are placed at the end of the frog rails as there is a possibility of shorts here.

In Fig. 1.16 the points P1 are set for 'straight through' and so we have a straight blue rail and a straight red rail. The red rail also goes as far as 'B', where it joins with the P2 closure rail which is yellow, meaning it is 'isolated'. Therefore the positive power only goes as far as B. The same happens with the negative on P1, as the closure rail is away from the bottom blue rail. As you can see, the upper stock rail is also red. The yellow rail shows the isolated rail from the lower blue rail.

Also the negative coming from P2 can only go as far as 'D'. So in this case there is no need to fit isolator fishplates because the points are doing the isolating.

In Fig. 1.17 the points P1 are set for 'turnout' so we have a straight blue rail and a blue 'closure' rail all the way to 'C'. On P2 the points are in 'turnout' so the blue rail connects at 'C–D': this means we have negative continuity. The red rail of P1 is on 'turnout' only and continues up to 'B'. On P2 the red rail continues to the junction of 'A–B' so we have positive power from track 1 to track 2.

This is all fine except you can only run the locomotive as far as the right-hand side of point P2, where it will derail if you continue. If you change the points to let it through you will cut power to track 2. The second problem is that you cannot get the loco off track 2 without reversing it. A simple answer to this problem is to fit continuity links to the points.

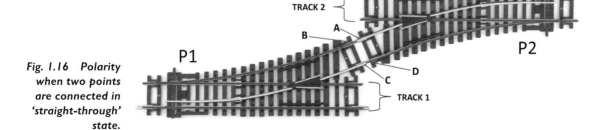

Fig. 1.16 Polarity when two points are connected in 'straight-through' state.

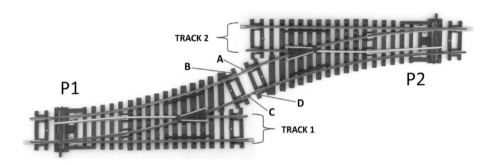

Fig. 1.17 Polarity when two points are connected in 'turnout' state.

CONTINUITY LINKS

These are stainless-steel spring links shaped as shown in yellow in Fig. 1.18. The image below shows the points in 'straight-through' (ST) position, with c1 closed and b1 open. However, due to the position of the links there is still positive power on rail 'c' and negative power on 'b'. This means that although the points are in ST position, any loco on rails b and d will work.

The image below (Fig. 1.19) shows the points in 'turnout' (TU) position, with c1 open and b1 closed. However, due to the position of the links there is still positive power on rail 'c' and negative power on 'b'. This means that although the points are in TU position, any loco on rails 'a' and 'c' will work.

By using continuity links on points P1 and P2 you have now allowed the feed and the return to go to all parts of the two tracks, so you will only need one speed controller.

The problem with this layout is that with one controller, if you run two locos they will both do the same speed and will both go into reverse if the polarity is changed. To run more than one loco individually you will need to have a separate controller on track 2 and isolate the point sets. So let's remove

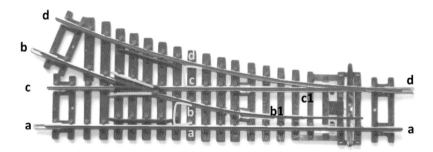

Fig. 1.18 Continuity clips on 'straight-through' points.

Fig. 1.19 Continuity clips on 'turnout' points.

THE LAYOUT

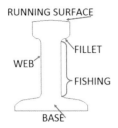

Fig. 1.20 Rail cross-section.

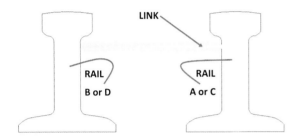

Fig. 1.21 Position of the continuity clip.

the continuity clips – though they still have a place on your layout, which we will come to later.

ADDING MORE POINTS

We have now added two more sets of points, P3 and P4, which gives the layout more versatility. In this layout one controller can run the entire layout, where connection is to both feeds and both returns. Now the position of the points doesn't matter, as power is on all tracks. Moving a train from one track to the next is possible by simply changing the points.

In this layout you can move the loco from track 2 to track 1 by changing points P4 and P3 to 'turnout'. Once on track 1, change P3 and P4 to 'straight through' and the loco will run around track 1. To take the loco back to track 2, change points P1 and P2 to 'turnout' and the loco will move back on to track 2. Once on track 2, change P1 and P2 to 'straight through'.

The limitations of this layout are that you can't control the speed of the two locos independently, and you can't stop one without stopping the other. The answer is to introduce another speed controller on track 2, making track 2 independent of track 1 – however, there is something you must take into consideration. Although the points are self-isolating (depending on the type you have purchased), you should fit isolating fishplates at the junction of P1 and P2, and also at the junction of P3 and P4, to prevent any possibility of a short circuit. If you are running track 1 in a clockwise motion and track 2 in a counter-clockwise motion, when you change any of the points you could create a dead short.

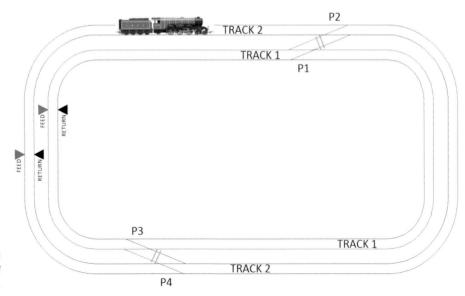

Fig. 1.22 Points P3 and P4 added to the layout.

ISOLATION FISHPLATES

Here we show the use of insulating rail joiners. They are the same as the metal ones except they are made in PVC, and they have a little block half way to keep the rails separated.

Note that if you have previously fitted isolation cancelling clips, you can leave them in, as they will now have no effect. However, good practice would be to remove them.

To recap: we now have two independently powered tracks where you can run two locos at different speeds and, if you desire, in different directions; also stopping and starting is independent.

MOVING LOCOS BETWEEN TRACKS

It is possible to move a loco from track 1 to track 2 and vice versa. The points need to be set to 'turnout' (TU) – though this will have no effect electrically, as you have fitted 'isolation fishplates' as shown above. You will need to ensure that both controllers are set at the same speed and direction, once the locos have changed tracks and the points have been set to 'straight through' (ST) – then you can adjust speeds independently for each loco. Alternatively you could leave the points at 'turnout'

Fig. 1.23 Position of the isolation fishplate.

(TU), and the locos will swap tracks for as long as you require.

PASSING LOOP

We are now going to run two trains on track 2. To do this we need a 'passing loop', which could be positioned at a station, or anywhere you require (see Fig. 1.25). A passing loop allows you to place a loco in the loop and isolate it from the main track, or to isolate the main track from the passing loop.

If we set P5 and P6 to 'turnout' (TU) position, train 3 will travel around track 2 and go into the passing loop at P6 and join track 2 again at P5. With the points in this position the main track 2 between P5 and P6 has no power to it, so train 2 cannot be moved. If you now run train 3 around the track and bring it back into the 'passing loop' and bring

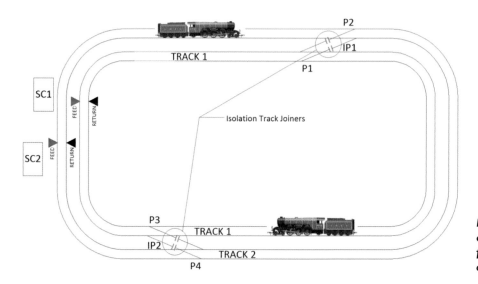

Fig. 1.24 Positions of the isolation fishplates in the complete layout.

THE LAYOUT | 23

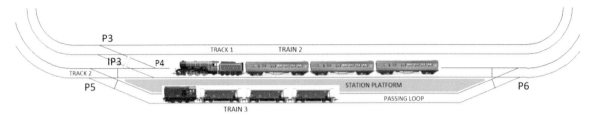

Fig. 1.25 The addition of points P5 and P6 to create a passing loop.

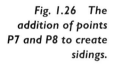

Fig. 1.26 The addition of points P7 and P8 to create sidings.

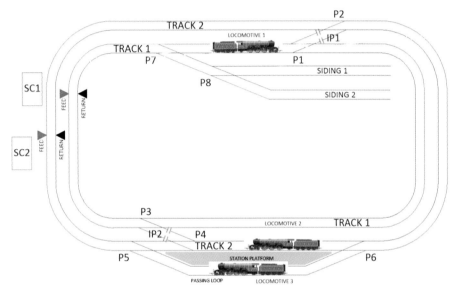

it to a standstill, you could change the points P5 and P6 back to 'straight through' (ST). This will remove power from the 'passing loop' and put it back on track 2 between P5 and P6. You can now take train 2 out of the station and run it round the track. This is all being controlled by a single speed controller.

SIDINGS

We are now going to introduce some sidings. These could be placed anywhere on track 1 or track 2; we have chosen to put one on track 1 in the position shown in Fig. 1.26. To create two sidings we need two sets of points, P7 and P8. If we now set P7 to 'turnout' (TU) and P8 to 'straight through' (ST) we have power to siding 2, so we could bring loco 1 around track 1 and into siding 2. Once the loco is in siding 2 we can change P8 to 'turnout' (TU), which will 'isolate' siding 2 from the power and put power on to siding 1.

With the arrangement set out above we can move all three locos to any position we require. For example: put loco 3 into siding 1. Set points P5 and P6 to 'turnout', bring loco 3 out of the passing loop and around to just past points P2. Change P1 and P2 to 'turnout' and P7 to 'straight through'. Set speed regulator C1 to reverse, and to approximately the same speed as C2. To move loco 3, set C2 to reverse, and increase the speed to match C1. The loco will now move backwards through P2 and P1, and continue backwards past P7. Stop, now change P7 to 'turnout' and P8 to 'turnout', set the direction to forwards, and bring the loco into siding 1, and stop.

With this layout the combinations of running locomotives is endless, and from here on you can

THE LAYOUT

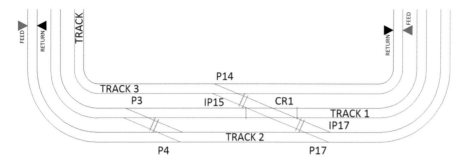

Fig. 1.27 A crossover in position CR1.

add more sections of track – just follow the principles set out above.

CROSSOVERS

You may want to introduce a crossover section into your layout. This is a section of track that crosses over another track, shown as CR1 in the drawing [Fig. 1.27]. There is nothing special about crossovers other than you need to know the electrical path created by the rails, which is shown in Fig. 1.28. As you can see, there are four distinct electrical paths, and because there are no moving parts these do not change physically; however, they do change electrically, depending on where the feed and return are connected. In the layout above it is still advisable to have isolation point IP15, because as you change the points P17 you will put power on to the crossover from track 2. Before you change P14 ensure that the power to the rails is the same as track 2, otherwise there will be a problem with the motion of the locomotive. It will not affect the electrical path because whatever the rails are on track 3, they are isolated from the crossover CR1.

The crossover is available in both left- and right-hand options. Fig. 1.28 shows the electrical paths on

Fig. 1.28 Rail polarity of a crossover.

the crossover. The main line runs from A to D, and as you can see, is the same polarity as in Fig. 1.27. The track B to C runs from track 2 to track 3, so B is connected to points P17, and C is connected to points P14.

SCISSOR CROSSOVER

A scissor crossover allows you to move from one track to the other in both directions; to do this you would normally need four sets of points. A scissor crossing also has four sets of point motors, one in each corner. In the following images we show the electrical paths including the frogs. The orange lines show the pieces of track that are isolated. In Fig. 1.29

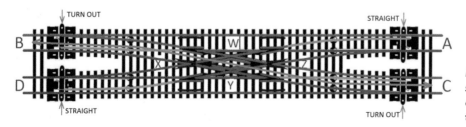

Fig. 1.29 Polarity of scissor crossover with all points straight through.

THE LAYOUT

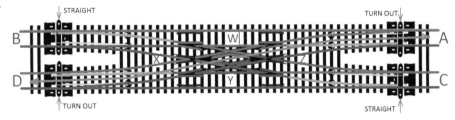

Fig. 1.30 Polarity of scissor crossover with two points in turnout (B and C).

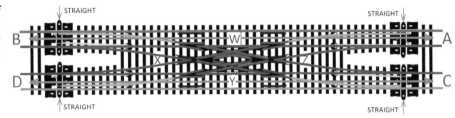

Fig. 1.31 Polarity of scissor crossover with two points in turnout (A and D).

the travel routes are A to B and C to D. The top track has red and blue rails: note the colours of the frogs w, x, y and z. The lower track has brown and green rails – again, note the colours of the frogs.

In Fig. 1.30 the travel is from B to C, or vice versa. Now the polarity of the top track must be the same as the bottom track, so if you have separate controllers for each track, ensure they are the same before switching the points. The polarity of the frogs has changed to suit the new route.

In Fig. 1.31 the travel is from A to D, or vice versa. Again the polarity of the top track must be the same as the bottom track, so if you have separate controllers for each track ensure they are the same before switching the points. The polarity of the frogs has changed to suit the new route.

Y OUTLET POINTS (DOUBLE SLIP)

'Y' points allow you to divide the track into two parallel tracks with the addition of a small curved section. Most manufacturers have several 'Y' junction points, and they are all around 11 to 12 degrees from the horizontal. Take a note of the polarity when the points are in the 'up' (B) position. The yellow shows the rails that are isolated.

When the points are in the 'down' position – that is, A to C travel – note the yellow isolation rail preventing power to the 'up' track.

The 'Y' points are wired in exactly the same way as a conventional set of points – see the wiring section.

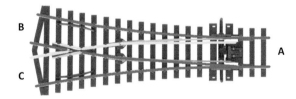

Fig. 1.32 'Y' outlet polarity in position B.

Fig. 1.33 'Y' outlet polarity in position C.

THREE OUTLET POINTS (TRIPLE SLIP)

The three outlet (triple slip) set of points is a little more complicated as it needs two sets of points, and unless it is near to you it is very difficult to see what state it is in. This is an ideal set to have indicators for all three outlets. Fig. 1.34 the travel is from A to B. You can see a red line running from the top rail of A to the top rail of B, and a corresponding blue rail running from the bottom rail of A to the bottom rail of B. Both C and D outlets have at least one yellow rail, so no power is getting to those tracks.

Fig. 1.35 the travel is from A to C. You can see a red line running from the top rail of A to the top rail of C, and a corresponding blue rail running from the bottom rail of A to the bottom rail of C. Both B and D outlets have at least one yellow rail, so no power is getting to those tracks.

Fig. 1.36 the travel is from A to D. You can see a red line running from the top rail of A to the top rail of D, and a corresponding blue rail running from the bottom rail of A to the bottom rail of D. Both B and C outlets have at least one yellow rail, so no power is getting to those tracks.

In Chapter 7, Track Wiring, we will show you how to wire up both a set of points, and give you an indication of how to show which line is in use at any time.

REVERSING LOOPS (Fig. 1.37)

This is a simple layout where the locomotive travels around two reversing loops, each positioned at the end of a section of track. The section between the loops does not have to be a straight track, it can have bends and so on. For simplicity we have drawn a straight track between the loops. There is one problem that needs to be overcome: if you follow the red rail (feed) from points P13 to the right, you will see it joins back at IP15 to the orange rail (Y), and this will cause a dead short. The solution is that the two loops have to be isolated from the connecting straight track at P12 and P13, and the feed between the loops has to change at certain times. This is why the indicators in the straight section are shown in orange and labelled X and Y.

For full wiring and switching, see Reversing Loops in Track Wiring.

Fig. 1.34   Three outlet points polarity in position B.

Fig. 1.35   Three outlet points polarity in position C.

Fig. 1.36   Three outlet points polarity in position D.

THE FIDDLE YARD (Fig. 1.38)

The fiddle yard is normally behind the scenery, hidden from public view, but is very much part of the layout. The design will be dictated by how you want to handle the stock – the yard can be completely manual, completely automatic, or somewhere in between. Each design has different space requirements which you must consider at the design stage. Fiddle yards can also contain a reversing loop for turning entire trains. There are several other types of fiddle yard, such as traverser, sector plate, cassette and elevator; however, for the purposes of this book we will concentrate on just the two.

THE LAYOUT

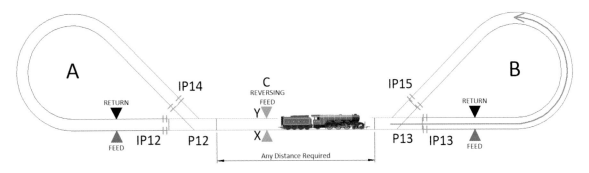

Fig. 1.37 Complete layout of a reversing loop.

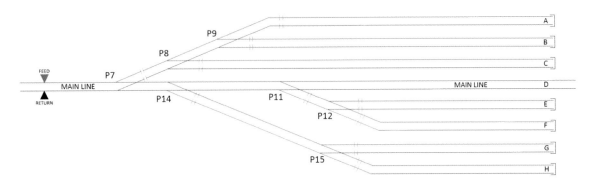

Fig. 1.38 Layout of a fiddle yard.

Broadly, designs can be in two categories. A fan of points giving many tracks can be easily constructed, operated automatically using point motors, and is simple to construct. The example in Fig. 1.38 is a fan-type fiddle yard. The main line runs through the middle with three separate positions to get to part of the fiddle yard.

The sidings has been given an identification letter from A to H. Each siding will need to have isolation points so that when the points are in the straight through (ST) position the siding is isolated from the power source. With the points P7, P14 and P11 in the straight through (ST) position, all the sidings are isolated by these points, so the loco will travel only on the main line.

THE TURNTABLE (Fig. 1.39)

The turntable is only used to store locomotives, as you can only put into the sidings a locomotive that fits on to the turntable bed, whereas the fan above can be used to store complete trains. Combinations of the two are also possible: for example, a fan could have a turntable off one of the fan outlets for storing locomotives separately from their trains, and allowing for them to be run reversed. The turntable must be completely isolated from all tracks using plastic insulator fishplates. Each siding and the turntable bed must have its own supply from the power supply via a reversing polarity switch, and an on/off switch. For a complete wiring diagram, see Chapter 7, Track Wiring.

STAGING YARD

A staging yard is again normally behind the scenery and is used to set up complete train assemblies of locomotive and carriages. This area would normally be where you manually pick and place a train assembly. Bringing them out on to the main layout would simply mean setting the correct points. For example,

THE LAYOUT

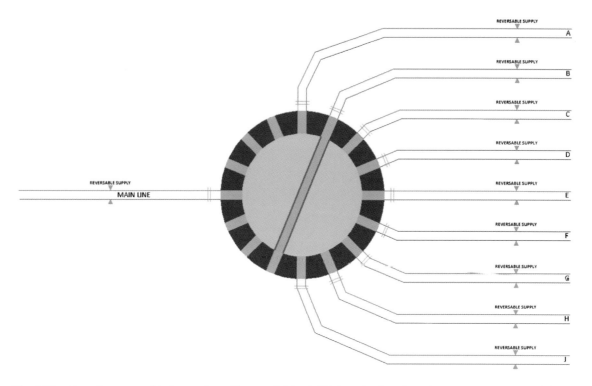

Fig. 1.39 Complete turntable layout (see Chapter 7, Track Wiring, for details on how to wire a turntable).

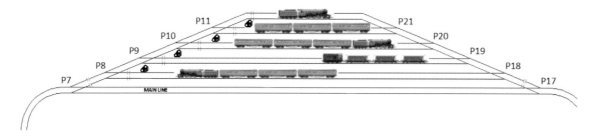

Fig. 1.40 Complete layout of a staging yard.

to bring out the goods train you would need to set points P7 and P9 to TU. This will put power from the main controller on to P9 track. Once the train is out on the main line set P7 to ST. To bring the goods train back into the staging area, set points P17 and P19 to TU.

Hopefully you now have a better knowledge of the component parts of your layout, and how it can be put together.

CHAPTER TWO

ELECTRONIC AND ELECTRICAL COMPONENTS EXPLAINED

Your model railway layout has a number of electronic components, and there may be times when you have to add components to get something to work properly. Here we try to explain to you what the component is, and how and why it works. *Please note* the information provided is by no means the full description and theory behind each component, but just enough to allow you to understand why it does what it does.

In this chapter we will build a simple power supply as we progress through each component, so you can see how and why it fits into the circuit. At the end of this chapter you will have a 2 amp 12v DC power supply, which can be used on any model rail layout.

TRANSFORMERS

A simple transformer consists of two electrical conductors called the 'primary winding' and the 'secondary winding', and a steel core that magnetically links them together. These two windings can be considered as a pair of mutually coupled coils. Energy is coupled between the windings by the magnetic field that links both primary and secondary windings. We know that a transformer generates its current output with the help of two windings, namely the primary and secondary windings. The primary coil of the transformer is always connected to the alternating power supply, and this is the only method of supplying power to the transformer, by connecting the power supply in parallel with the two free ends of the primary windings. The current produced is then transferred to the secondary winding by Faraday's law of mutual induction.

> In other words, on the whole it reduces the input voltage to a specified output voltage without the input having any contact with the output

Fig. 2.2 shows a transformer with one secondary output, and Fig. 2.3 shows a transformer with two secondary outputs.

A transformer can have more than one primary winding and more than one secondary winding, so if two or more coils exist at any terminal, then they can be connected with each other in two basic ways. The two ways for connecting the two or more windings with each other are described below.

SERIES CONNECTION OF WINDINGS

The secondary windings of a transformer, connected in series, are shown in Fig. 2.4.

Fig. 2.1 *The circuit diagram for a transformer, and a few examples of the physical units.*
JPR ELECTRONICS

30 ELECTRONIC AND ELECTRICAL COMPONENTS EXPLAINED

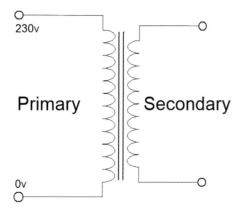

Fig. 2.2 *Single secondary output.*

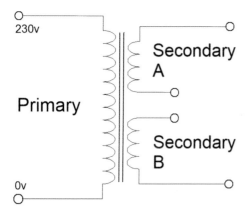

Fig. 2.3 *Dual secondary output.*

Now at the secondary terminal, if we are using two windings rated at 12v each, as shown in Fig. 2.4, then we get a total of 12v + 12v = 24v at the output terminal, which means that the voltage has added up. The same amount of current will flow through each of the windings, which is 1A for each in this case, and hence the total current at the output is also equal to 1A.

From this we can conclude that, if we want to get twice the voltage at the output, we can connect two secondary windings in series and so on, provided that the current remains constant.

PARALLEL CONNECTION OF WINDINGS

The secondary windings of a transformer, connected in parallel, are shown in Fig. 2.5.

As we know that the current in a parallel combination adds up, in this case we see that both the secondary windings that have been connected in parallel are rated at 1A each, so the current will add up here, and as there are two windings, so it adds up as 1A + 1A = 2A in total. Since the voltage remains the same, so the voltage drop across the output terminal will be the same as that on each of the windings of the transformer, and the output will be rated at 12v, 2A transformer.

If we use a dual voltage transformer, then only the readings of the current and voltage will be changed accordingly, but the principle that the voltage remains

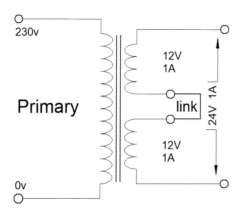

Fig. 2.4 *Secondary output connected in series.*

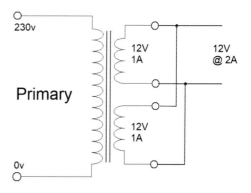

Fig. 2.5 *Secondary output connected in parallel.*

constant in parallel combination and the current adds up at the output, remains the same. Moreover, other than these two configurations, other types of connection are also available, such as driving two independent outputs by two individual secondary coils of a transformer, such that both of them have no interconnection.

One thing that should be taken care of is that while connecting two windings with each other, their phase relationships should be kept in mind and the connections should be made accordingly. If the terminals with opposite phase relationships are connected with each other, they will cancel the effect of each other's magnetic flux, and hence we won't get any output. So in order to get the desired output, only the terminals with the same phase relationships should be connected with each other.

The power supply that we will build through this chapter starts with the transformer (Fig. 2.6).

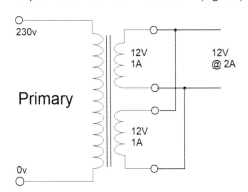

Fig. 2.6 *Let's build our own power supply through this chapter.*

DIODES

A diode is an electrical device allowing current to move through it in one direction with far greater ease than in the other. The most common kind of diode in modern circuit design is the semi-conductor diode, although other diode technologies exist.

> In other words, it allows current to flow forwards (from anode to cathode) but not backwards (cathode to anode)

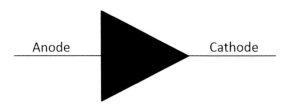

Fig. 2.7 *Schematic symbol for a diode.*

Fig. 2.8 *A physical diode.* JPR ELECTRONICS

Fig. 2.7 shows the schematic symbol for a diode, and Fig. 2.8 the real life appearance of a diode. A diode always has a line (white or black depending on the background) to denote the cathode.

SIGNAL DIODE (Fig. 2.9)

As mentioned above, the signal diode is very useful to block certain voltages from other voltages. The best example of this is the simple circuit in Fig. 2.9. With the switch in position 'A', the red and yellow lamps are 'on' and the green is 'off'. In position 'B', the yellow and green are 'on' and the red is 'off'. The reason for this is that in position 'A', diode D2 is blocking the voltage getting to the green lamp, and in position 'B' the diode D1 is blocking the voltage getting to the red lamp.

BRIDGE RECTIFIER (Fig. 2.10)

Diodes are also used to convert AC (alternating current) into DC (direct current). This can be done by using four power diodes, or a pre-made bridge rectifier. Fig. 2.10 is the schematic symbol for a bridge rectifier, and an example of the physical units. The diodes always have the positive terminal marked on the casing.

AC, or alternating current, means that the electrical current is alternating direction in a repetitive pattern. The frequency of repetition of this current is 60 Hertz; this means that the direction of the

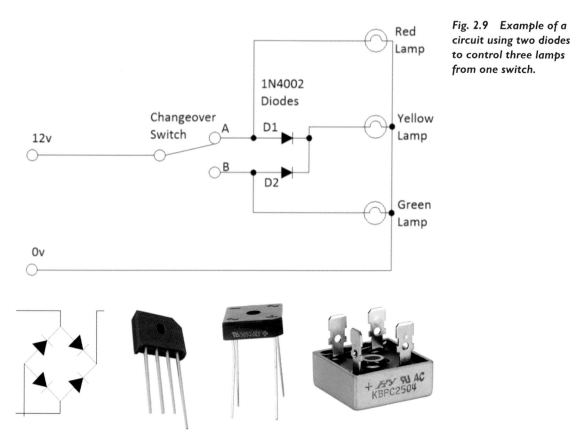

Fig. 2.9 Example of a circuit using two diodes to control three lamps from one switch.

Fig. 2.10 The circuit diagram for a bridge rectifier and a few examples of the physical units.
JPR ELECTRONICS

current changes sixty times every second. This is the type of current delivered to your home or business. See graph (Fig. 2.11) below.

The bridge rectifier only allows the positive cycles to pass through the diodes, so at one end of the rectifier you have 0v and at the other end you have +v. See graph (Fig. 2.12) below.

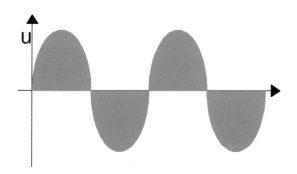

Fig. 2.11 Output from the transformer before the bridge rectifier, which is AC.

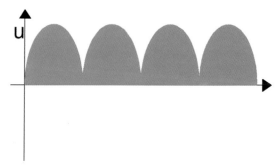

Fig. 2.12 Output from the bridge rectifier once connected to the transformer, showing all the lower voltages clipped.

ELECTRONIC AND ELECTRICAL COMPONENTS EXPLAINED | 33

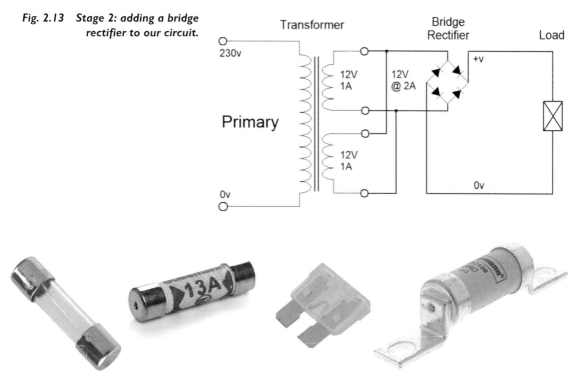

Fig. 2.13 Stage 2: adding a bridge rectifier to our circuit.

Fig. 2.14 A range of fuses. JPR ELECTRONICS

Our power supply now has a bridge rectifier. The AC comes from the plug in your home, through a transformer into the bridge rectifier. This gives an output of DC for your load.

FUSES

A fuse is an automatic means of removing power from a faulty system: it is made up of a metal wire in a housing that fits into a fuse holder; the metal strip has a predetermined current carrying capability which when exceeded, melts the wire and disrupts the flow of electricity through it. It is a sacrificial device, unlike a circuit breaker, which will reset and can be used again and again.

There are thousands of different fuse designs which have specific current and voltage ratings, breaking capacity and response times, depending on the application. The time and current operating characteristics of fuses are chosen to provide adequate protection without needless interruption.

Fuses come in different physical sizes and different current ratings, and these can be further divided into response times such as quick blow, slow blow, time delay and anti-surge. In the main you will only be concerned with basic domestic fuses, the sort that fit in the plug, and quick blow fuses.

The criterion for selecting a fuse is very simple: it must protect the equipment before the fuse. Therefore a fuse in the domestic plug (although called a 13-amp plug) should be rated at the top end of the equipment: that is, if you have a 5-amp power supply the fuse should be 5 amps in the household plug. Any higher than the rating of the power supply and it will affect the household supply before the fuse blows.

The fuse must protect the power supply at its output. If you are running a simple timer circuit from a power supply that draws ½ an amp (500mA), then the fuse should be no more than 500mA, because if the timer starts to draw more than 500mA there is

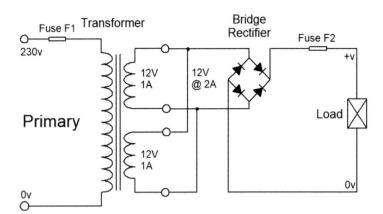

Fig. 2.15 Stage 3: adding a fuse to our circuit.

clearly a fault with the timer and you need to switch it off.

The problem arises when you are running a number of items from the same power supply and all add up to, say, 2 amps, then clearly the 500mA fuse will not cope, so a 2-amp fuse should be fitted. In this case the prudent thing to do is to have each piece of equipment fused with the correct fuse, so that if one develops a fault it blows its own fuse and not the main power supply fuse, and all the other equipment remains running.

In our circuit we have now added two fuses: F1 protects the household supply, and F2 protects the power supply.

CIRCUIT BREAKER

A circuit breaker is like a fuse as it protects equipment when a short or an overload is detected. In the case of a fuse it will blow when the current goes above that of the fuse rating, whereas a circuit breaker will open a contact (switching power off), and then will reset when the current is below its rating or the short has been fixed.

The first drawing in Fig. 2.16 is the schematic symbol for a circuit breaker; and examples of the physical units are also shown. As it is essentially an on–off switch there is no polarity to it.

How does this happen: In a circuit breaker the switch is made of a bimetal strip (two metals fused together). The two metals are selected because they bend at different rates when heated. As the current rises, the bimetal strip heats up; when the current is above the switch rating the bimetal strip will have heated enough to bend it away from the other contact – this then opens the switch so no power can flow through it. As there is no power going through the switch, the bimetal strip starts to cool down and bends back into shape; it will make contact with the other part of the switch, which then connects the power back to the circuit it is protecting. Most modern equipment will have one of these in the circuit somewhere.

Where not to use a circuit breaker: As you can see from the circuit in Fig. 2.17 the breaker has been used right at the beginning of the power supply. Circuit breakers are slower than some fuses to switch off, due to the time it takes to heat up the bimetal strip, so in the case of sensitive electronic

Fig. 2.16 Circuit diagram and images of circuit breakers.
JPR ELECTRONICS

ELECTRONIC AND ELECTRICAL COMPONENTS EXPLAINED

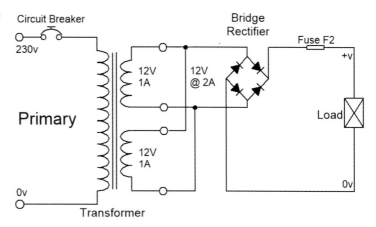

Fig. 2.17 Stage 4: adding a circuit breaker to our circuit.

equipment a fuse F2 must be used. (Please note that in all respects we are talking milli seconds, but that is enough to damage an electronic circuit.)

In Fig. 2.17 the fuse that was before the transformer in Fig. 2.16 has been replaced with a circuit breaker CB1. The current rating of the circuit breaker is set at just below, or the same as, the transformer rating.

ZENER DIODE

A zener diode allows current to flow from its anode to its cathode like a normal semiconductor diode, but it also permits current to flow in the reverse direction when its 'zener voltage' is reached. The zener still conducts electricity in a forward direction, like any other diode, but it also conducts in the reverse direction, if the voltage applied is reversed and larger than the zener breakdown voltage.

> In other words, it allows voltage to pass the diode up to the value of the zener diode and no more

A zener diode always has a line (white or black depending on the background) to denote the cathode.

In the circuit shown in Fig. 2.20 the zener diode is controlling the output voltage at vout. If the vin voltage is 12vDC and the zener Vz is 9v, then the

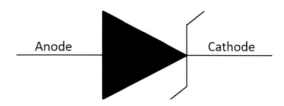

Fig. 2.18 Schematic of a zener diode.

Fig. 2.19 An actual zener diode. JPR ELECTRONICS

output at vout will be 9v. The resistor would be 3.9 ohms rated at 2.3 watts, and will give an output of 9v at 500mAmps.

The formula is resistor value (R) = (vin–vout) / I, where 'I' is the current required at the output.

We are now getting to the stage where we have more control over the output from the transformer.

ELECTROLYTIC CAPACITORS

All electrolytic capacitors are polarized capacitors whose anode (+) is made of a particular metal on which an insulating oxide layer forms by anodization, acting as the dielectric of the electrolytic capacitor. A non-solid or solid electrolyte that covers the surface

36 ELECTRONIC AND ELECTRICAL COMPONENTS EXPLAINED

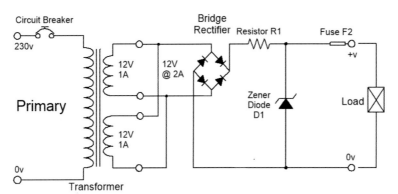

Fig. 2.20 Stage 5: adding a zener diode to our circuit.

of the oxide layer in principle serves as the second electrode (cathode) (−) of the capacitor.

> In other words, a capacitor stores and releases power (a little like a battery). The amount it stores is dependent on the size, and the rate at which it dissipates it is dependent on the subsequent load

Electrolytic capacitors are available in 'axial' (a lead at each end) or 'radial' (both leads at one end) form. The casing will show the negative lead, which with radials will be shorter than the positive.

Electrolytic capacitors are used in power supply units to smooth out the output voltage. If there are any ripples in the voltage from the transformer the capacitor smooths them out because it gives out stored power during a ripple. (This is a very simplistic description, and there is more to a smooth power supply than just putting a capacitor across the output.)

In the circuit depicted in Fig. 2.22, the capacitor is placed after the zener diode to smooth the output – it could have been put before the resistor R1 to smooth the output from the bridge rectifier (this we do later on).

Fig. 2.21 Circuit diagram and images of electrolytic capacitors.
JPR ELECTRONICS

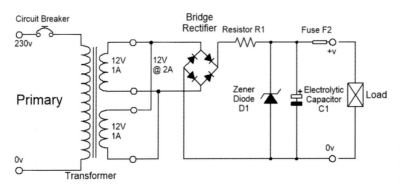

Fig. 2.22 Stage 6: adding an electrolytic capacitor to our circuit.

ELECTRONIC AND ELECTRICAL COMPONENTS EXPLAINED

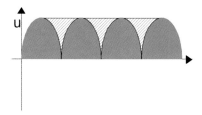

Fig. 2.23 Before adding the capacitor.

Fig. 2.24 After adding the capacitor.

The capacitor will try to fill the gaps (the hatched area in Fig. 2.23) in the power output. It does this because it holds the 12v while the wave goes to 0v. Without this the output would be like a pulse of 12v. It has to be said this pulse is so quick you wouldn't notice it, but some electronic circuits don't like it. So with the capacitor in place the output should look more as shown in Fig. 2.24.

(Note: Care should be taken with used larger capacitors: they have been known to hold electrical power for a long time, and can give you quite a shock.)

LIGHT-EMITTING DIODE (LED)

A light-emitting diode (LED) is a two-lead semiconductor light source. It is a p–n junction diode, which emits light when activated. When a suitable voltage is applied to the leads, electrons are able to recombine with electron holes within the device, releasing energy in the form of photons.

> In other words, when current is passed through the LED from anode to cathode the LED will be illuminated; if the current/voltage is reversed, nothing will happen

An LED always has a flat on the body and a shorter lead to denote the cathode (C). The circuit diagram illustrated looks like a diode with arrows to show that it emits light.

LEDs only require a small voltage to make them work, so when using them on a 3- to 24-volt DC circuit you will require a 'dropping resistor' to bring the voltage down to the working voltage of the LED. Note that some LEDs have built-in resistors, so will state in their description their working voltage – that is, 12vDC.

The formula to use is resistor (R) = (voltage supply–LED voltage) / LED current.

In the circuit in Fig. 2.26 the LED is used to show that the power in 'on'. The negative side of the LED is connected to the 0v rail, and the positive side of the LED is connected to a dropping resistor, which is connected to the positive rail of the power supply. Using the formula above will determine the value of the dropping resistor.

FILAMENT LAMPS

A filament lamp is a common type of light bulb. It contains a thin coil of wire called the filament. This

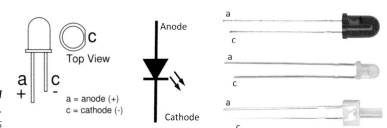

Fig. 2.25 Schematic diagram and images of LEDs.
JPR ELECTRONICS

38 ELECTRONIC AND ELECTRICAL COMPONENTS EXPLAINED

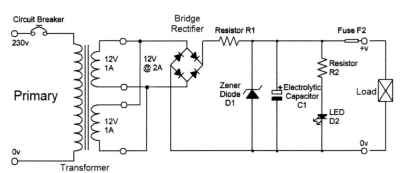

Fig. 2.26 Stage 7: adding an LED to our circuit to show when it is 'on'.

Fig. 2.27 Schematic diagram and images of filament lamps.
JPR ELECTRONICS

heats up when an electric current passes through it, and produces light as a result. The resistance of a lamp increases as the temperature of its filament increases. Most of the air is removed from the bulb and replaced with an oxygen-free gas (inert) to prevent the filament from oxidizing (burning up) when it gets hot and glows.

There are a number of words you will hear when talking about filament lamps:

Candela (cd): A unit of luminous intensity in the International System of Units (SI), defined as the luminous intensity in a given direction of a source that emits monochromatic radiation of frequency 540×10^{12} hertz, and has a radiant intensity in that same direction of $1/683$ watt per steradian (unit solid angle).

Lumens: The higher the lumen rating, the brighter the lamp will appear. It is a more accurate measure than watts for denoting the brightness of a light bulb. So when looking at filament bulbs, the brightness is directly related to the amount of current it draws. This has to be taken into consideration regarding the output current from your power supply.

In the circuit illustrated in Fig. 2.28 we have replaced the LED with a filament lamp, LA1.

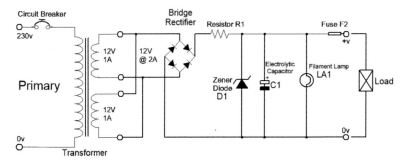

Fig. 2.28 Stage 8: we have replaced the LED with a 12v filament lamp, if you prefer this to an LED.

ELECTRONIC AND ELECTRICAL COMPONENTS EXPLAINED

RESISTORS

A resistor is a passive two-terminal electrical component that implements electrical resistance as a circuit element. Resistors may be used to reduce current flow, and at the same time, may act to lower voltage levels within circuits. In electronic circuits, resistors are used to limit current flow, to adjust signal levels.

> In other words, a resistor reduces the amount of power coming out of it

A resistor will have a series of coloured strips that tell you its value, its tolerance, and sometimes the wattage.

There are many different types and shapes of resistor, but for your purposes a carbon film resistor is adequate. These are available in values from 1 ohm to 10m ohm, and in a different power rating (wattage).

RESISTOR COLOUR CODES

The most common types of resistor have four colour bands. Band one gives you the first digit of the value, band two gives the second digit of the value, and the third band gives the multiplier. In the example of the resistor colour code table in Fig. 2.30, brown is a one,

Fig. 2.29 Schematic diagram and images of fixed value resistors. JPR ELECTRONICS

4 Band Resistors

120K RESISTOR with
4th band Silver = 10% Tolerance
4th band Gold = 5% Tolerance

BAND 1	BAND 2	BAND 3							
		GOLD	**BLACK**	**BROWN**	**RED**	**ORANGE**	**YELLOW**	**GREEN**	**BLUE**
BROWN	BLACK	1R0	10R	100R	1K0	10K	100K	1M0	10M
BROWN	RED	1R2	12R	120R	1K2	12K	120K	1M2	12M
BROWN	GREEN	1R5	15R	150R	1K5	15K	150K	1M5	15M
BROWN	GREY	1R8	18R	180R	1K8	18K	180K	1M8	18M
RED	RED	2R2	22R	220R	2K2	22K	220K	2M2	
RED	VIOLET	2R7	27R	270R	2K7	27K	270K	2M7	
ORANGE	ORANGE	3R3	33R	330R	3K3	33K	330K	3M3	
ORANGE	WHITE	3R9	39R	390R	3K9	39K	390K	3M9	
YELLOW	VIOLET	4R7	47R	470R	4K7	47K	370K	4M7	
GREEN	BLUE	5R6	56R	560R	5K6	56K	560K	5M6	
BLUE	GREY	6R8	68R	680R	6K8	68K	680K	6M8	
GREY	RED	8R2	82R	820R	8K2	82K	820K	8K2	

Fig. 2.30 Resistor colour code table to calculate the value of a resistor. BRIMAL

40 ELECTRONIC AND ELECTRICAL COMPONENTS EXPLAINED

red is a two, and yellow is ×10, giving a resistance of 120K; the final band is silver, which gives that value a tolerance of 10 per cent.

In the circuit shown in Fig. 2.31 there are two resistors, both with similar functions. R1 is designed to reduce the voltage for the zener diode, and because the power supply is rated at 500mA the current flowing through this resistor will be high – it has a lot of work to do, and the wattage of the resistor needs to be high otherwise it will overheat and eventually break down.

R2 is designed to reduce the voltage to the LED – see the formula for this above. As the LED does not draw very much current, the wattage of this resistor can be lower; however, again, if too low it will overheat, with the same result.

The value of the resistor has nothing to do with the wattage (simplified): thus R1 has a high wattage but a lower resistance value than R2, which has a lower wattage rating but a higher resistance value.

PRE-SET RESISTOR

A pre-set is a three-legged electronic component which can be made to offer varying resistance in a circuit. The resistance is varied by adjusting the rotary control on the pre-set resistor. The adjustment can be done by using a small screwdriver or a similar tool. The resistance does not vary linearly, but rather varies in exponential or logarithmic manner. Such variable resistors are commonly used to adjust the length a timer is on. The variable resistance is obtained across the single terminal at the front, and one of the two other terminals. The two legs at the back offer a fixed resistance, which is divided by the front leg. So whenever only the back terminals are used, a pre-set acts as a fixed resistor. Pre-sets are specified by their fixed value resistance.

In our circuit we have now added a voltage regulator RG1. This is designed to regulate the

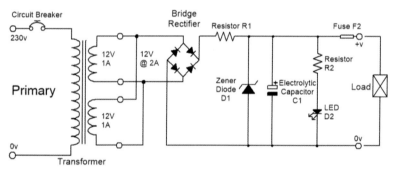

Fig. 2.31 Stage 9 shows the position of two resistors, R1 and R2: see text for explanations.

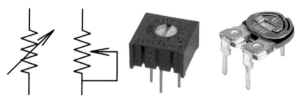

Fig. 2.32 Schematic diagram and images of pre-set (variable) resistors. JPR ELECTRONICS

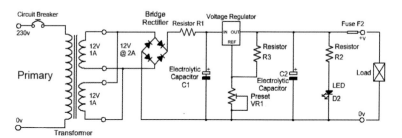

Fig. 2.33 Stage 10 shows the position of the pre-set resistor VR1.

ELECTRONIC AND ELECTRICAL COMPONENTS EXPLAINED

> In other words a pre-set resistor reduces the amount of power coming out, which is variable, from none to full – a little like a water tap

output voltage more precisely. To do this we have added a pre-set resistor VR1. This pre-set resistor adjusts the output voltage to give you exactly the voltage you require in the output.

VOLTAGE REGULATOR

A voltage regulator is any electrical or electronic device that maintains the voltage of a power source within acceptable limits. The voltage regulator is needed to keep voltages within the prescribed range that can be tolerated by the electrical equipment using that voltage.

If the reference (ref) pin is connected to ground (0v), the output pin delivers a regulated voltage of 1.25v at currents up to the maximum. Higher regulated voltages are obtained by connecting the reference pin to a resistive voltage divider between the output and ground (0v). In the circuit in Fig. 2.33, the resistive voltage divider is created by VR1 and R3: this allows the output voltage to be varied from 1.2v to the maximum of the transformer.

A voltage regulator is a complex circuit with transistors, resistors and diodes all miniaturized and put into a small package, as shown in Fig. 2.35.

The voltage regulator generates a great deal of heat – in some cases it becomes too hot to touch. To keep this heat under control and reduce it, the regulator would normally be mounted on a heatsink.

Fig. 2.34 Schematic diagram and images of voltage regulator integrated circuits.
JPR ELECTRONICS

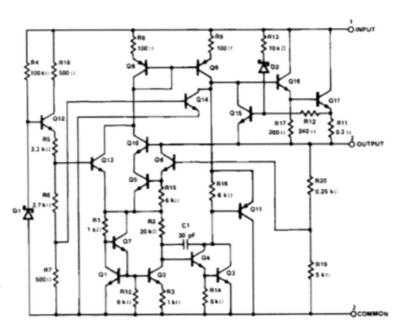

Fig. 2.35 The complex circuitry inside a voltage regulator integrated circuit.
JPR ELECTRONICS

42 | ELECTRONIC AND ELECTRICAL COMPONENTS EXPLAINED

Fig. 2.36 **Heatsinks for the voltage regulator.** JPR ELECTRONICS

For this project a heatsink that has a thermal resistance of 17°C/W would be adequate. Thermal resistance is a heat property, and a measurement of a temperature difference by which an object or material resists a heat flow.

RELAYS

A relay is an electrically operated switch. Relays use an electromagnet to operate a switch mechanically, but other operating principles are also used, such as solid-state relays. Relays can be energized by a low power signal.

A simple electromagnetic relay consists of a coil of wire wrapped around an iron core (a solenoid); when an electric current is passed through the coil it generates a magnetic field that activates the armature, and the consequent movement of the movable contact(s) either makes or breaks (depending upon construction) a connection with a fixed contact.

The circuit diagram of a relay is shown as a box with a wire at the top and the bottom: this is the coil. Next to it are the contacts, which could be one or more.

The purpose of a relay is to isolate one section of a circuit from another, and also to switch 'on' or 'off' a higher power than that of the circuit controlling it – thus a 12v DC relay could be used to switch a 10-amp 240v AC motor on and off.

Relays come in many forms and contact arrangements: some fit on to printed circuit boards, some fit into a socket for chassis mounting. They also come with a range of coil voltages, from 5v DC to 240v AC. The contact arrangements vary from normally open (closes when energizes) to changeover (changes from one contact to another). (See Appendix II for further details.)

We now add a relay to our power supply because we want to switch another power source of a different voltage on and off. The contact arrangement of the relay shown is double pole changeover. The

Fig. 2.37 **Schematic diagram and images of miniature relays.** BRIMAL/JPR ELECTRONICS

ELECTRONIC AND ELECTRICAL COMPONENTS EXPLAINED

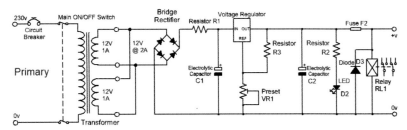

Fig. 2.38 Stage 11: the introduction of a relay on the output of our power supply circuit. BRIMAL

position of fuse F2 could be either before the relay, as shown, or after the relay. In the circuit shown in Fig. 2.38 the relay is being considered as part of the load. The + and 0 outputs can be used to power equipment that is rated at 12v DC.

EMF (ELECTRO MOTIVE FORCE)

The collapsing magnetic field, when an inductive circuit is broken, causes a voltage transient or back EMF. This will eventually lead to complete welding of the switch. Coil-operated relays or solenoid control valves possess considerable inductive values. The significant factor to guard against is the release of energy, temporarily stored in the coil, at the time the switch opens.

The switch can be of any type where two metals join together to make a circuit: reed switch, toggle switch, leaf switch, push-button switch, relay, and so on.

The diode is excellent for removing the inductive voltage spike, as the back EMF is directed through the diode. Contact erosion is reduced to a minimum, but when placed across a relay coil, the release time of the relay contacts will be increased by several milliseconds while the stored energy is being dissipated. The diode chosen must have a forward current rating equal to, or greater than, the steady state current of the circuit, and the diode must be connected cathode to positive.

SWITCHES

We now have a fully functional power supply unit; however, we have no way of switching it on or off. We need to introduce a switch just before the transformer. The switch needs to be a double pole on/off switch, it needs to be able to switch 230v AC, and at a current rating in excess of 2 amps. (See Appendix II for more details on switches.)

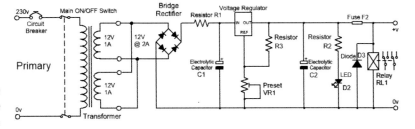

Fig. 2.39 Stage 12: adding a diode D3 across the coil of the relay to reduce EMF.

Fig. 2.40 Schematic diagram and images of 'on' and 'off' switches.
JPR ELECTRONICS

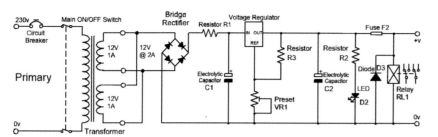

Fig. 2.41 The introduction of an 'on'/'off' switch for the power supply.

In the final circuit diagram shown in Fig. 2.41 we have added a main on–off switch to complete the project.

In Conclusion

While explaining the function of each component we have made a usable power supply unit. The LED D2 may stay on for a few seconds after you switch off the PSU. This is because the capacitor C1 is holding a charge, for reasons explained above, but it will eventually dissipate and the LED will go out. Each of the components may be used on your layout in isolation. Resistors will be used to reduce voltage for LEDs, which you may use for control panel indicators. Capacitors will be used to extend the indication for track occupancy circuits.

CHAPTER THREE

ELECTRICAL SYSTEMS AND POWER SUPPLIES

Most layouts require three forms of power source. The first is a variable 12v supply to control the speed of the train, the second is a fixed 12v supply to power all ancillary equipment such as points and signals, and the third is 16v AC for some types of points and signals.

So how much power do you need? Electrical power is a function of voltage and current (literally: power in watts is equal to the current in amps multiplied by the voltage in volts, or P=IV). Power for model trains and accessories is typically measured in thousandths of an amp, or milliamps (mA). Note: one milliamp of current at 12 volts provides 12 milli Watts of electrical power.

What really matters in a model train is the torque produced by the motor (which varies with current) and the power to move a train ('power in rotational motion', which is at its maximum at about half the full unloaded speed). Torque is used to overcome friction, and is particularly important when starting a stationary train: this we will cover in Chapter 5 Control Panels, but you need the power available to run the speed controller.

THE POWER SOURCE

There are many types of power supply unit (PSU) on the market today. What you are looking for is a 12v DC supply with a current rating of between 2 to 5 amps. The current rating required will depend on the size of your layout, and how many locomotives you are running at the same time; for very large layouts you could go up to 10 amps or higher. You will also need a 16v AC supply if you are using certain types of point motors and signals.

The first example is a power supply delivering 12 volts DC and rated at 3 amps.

The second example is a PSU delivering 12 volts and rated at 5 amps.

The third example is a PSU delivering 9 to 15 volts and rated at 10 amps. This unit has a potentiometer for you to set the voltage you require.

The fourth example is a 16v AC power unit with a separate circuit to convert the 16v AC to 12v DC at 1.1 amp, if required.

Fig. 3.1 Example of a power supply with a fixed output of 12v, rated at 3 amps. CPC

Fig. 3.2 Example of a power supply with a variable output from 3v to 12v DC, rated at 5 amps.
JPR ELECTRONICS

Fig. 3.3 Example of a power supply with an adjustable output from 9v to 15v, rated at 10 amps.
BRIMAL

46 ELECTRICAL SYSTEMS AND POWER SUPPLIES

Fig. 3.4 Example of a 16v AC power unit with an independent circuit to convert the output to 12v DC at 1.1 amp. GAUGEMASTER

Alternatively you could use the PSU you have just built in Chapter 2, Electronic and Electrical Components Explained.

Whichever one you decide to use, you will now have to distribute the power to certain equipment on your layout: the diagram in Fig. 3.5 gives an idea of how this could be done. You must, however, bear in mind the current capabilities of the PSU. The fourth example would appear to be the logical one, as you have the 16v AC and the 12v DC all in the one unit. However, the 12v DC is only rated at 1.1 amp, so will be limited as to what it can power. The circuit diagram for the fourth example is shown in Fig. 3.5.

A better power supply system is to use the 16v AC power pack just for the AC supply, and introduce a high current power unit for everything else; then you have enough power to run a number of speed controllers and all the other 12v DC equipment.

> In all drawings blue is shown as AC, red is positive (+) DC, and black is negative (0) DC

Now that the power is sorted out, we need to look at the speed controller.

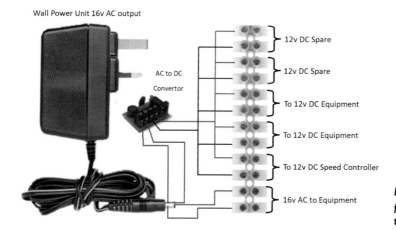

Fig. 3.5 Basic layout of a power unit with take-off terminals.

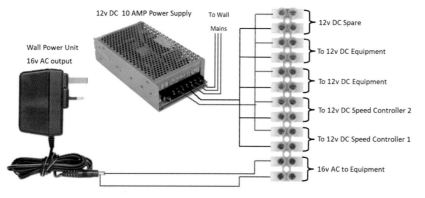

Fig. 3.6 A better power supply unit giving both AC output and DC outputs at a higher ampage.

ELECTRICAL SYSTEMS AND POWER SUPPLIES

SPEED CONTROLLERS

Most proprietary manufacturers make their own speed controller – a few examples are shown below (note that this is just a small sample). They all supply a variable voltage to the tracks, and most have a forward/reverse switch to change the polarity on the tracks. Some speed controllers will be supplied with the power unit built in so you may not require a power supply unit. However, these will be limited to the amount of current they can deliver, so at some stage you may have to invest in a separate PSU.

Hornby R8250: Controls the direction and speed of one locomotive. Input is 16v AC from the standard wall-plug mains transformer (P9000). Output is 0–12v to the track. It is supplied fitted with a 760mm lead with plug, to connect to the R8206 power track. This item does not include a transformer.

Kato 22-015 controller: A single-track controller designed for use with the Kato track system; however, it can be used with other track systems. This controller features a two-pin plug on the back, which allows the feeder wire (included) to be plugged in without the need for a screwdriver. It also features a male and female plug on the side that supplies 12v DC, to allow for accessories.

Gaugemaster Model D: Twin track transformer controller for 'N', 'OO', and most small gauges, with two accessory outputs.

Input: Mains (240 volts AC); use a 3-amp fuse
Output: 2 × 12v DC at 1 amp, each controlled
Accessories: 1 × 16v AC at 1 amp uncontrolled
　　　　　　 1 × 12v DC at 1 amp uncontrolled

Fig. 3.10: This is a basic PCB speed control module of motors up to 3 amps maximum current consump-

Fig. 3.7 A Hornby R8250 single-track speed controller. HORNBY

Fig. 3.9 A Gaugemaster dual-track speed controller. GAUGEMASTER

Fig. 3.8 A Kato 22-015 single-track speed controller. KATO

Fig. 3.10 An uncased speed controller can be mounted in your own console. JPR ELECTRONICS

48 ELECTRICAL SYSTEMS AND POWER SUPPLIES

tion. Pulse-width modulation (see below) provides proportional output control and torque maintenance. Output control is achieved by varying the output pulse.

We now have the basis for the power system for the layout. The wiring of these is straightforward, and all will follow the circuit as shown in Fig. 3.11.

> In all drawings blue is shown as AC, red is positive (+) DC, and black is negative (0) DC

As your system becomes larger and you put more equipment on the layout you may find that some of these items only require 5v DC. This is ideal for powering LEDs, and as you fit more it would be a good idea to have a separate power supply unit for them; this will be a 5v DC, and is available as a plug-in unit.

PULSE-WIDTH MODULATION

While looking for a speed controller you will come across the phrase 'pulse-width modulation' (PWM).

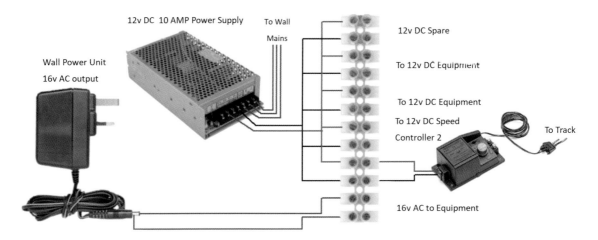

Fig. 3.11 A power layout with a speed controller attached.

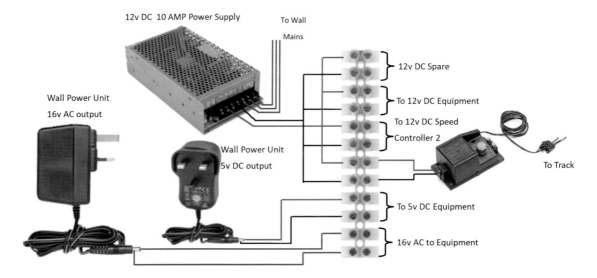

Fig. 3.12 A power layout with an additional 5v power pack for all low voltage circuits.

This is a modulation technique used to convert a voltage into a pulsing signal; it allows the control of the power supplied to electrical devices, especially to loads such as motors. The pulse-width modulation speed control works by driving the motor with a series of 'on–off' pulses, and varying the duty cycle – the fraction of time that the output voltage is 'on' compared to when it is 'off' – of the pulses, while keeping the frequency constant.

The power applied to the motor can be controlled by varying the width of these applied pulses, and thereby varying the average DC voltage applied to the motor's terminals. By changing or modulating the timing of these pulses the speed of the motor can be controlled – that is, the longer the pulse is 'on', the faster the motor will rotate, and likewise, the shorter the pulse is 'on', the more slowly the motor will rotate.

Also the amplitude of the motor voltage remains constant so the motor is always at full strength. The result is that the motor can be rotated much more slowly without it stalling.

ADD AN AMMETER

As mentioned before, the voltage is not the problem with a layout, it is the current drawn. As your layout grows, so does the demand for power, whether it be your 12v DC, 5v DC, or 16v AC supply. The current you draw may need to be monitored if you are expanding. A panel-mounted 10-amp meter is readily available and is connected into the positive wire as near to the power supply unit as possible. Where the positive wire goes into the terminal block, remove it and connect it to one of the terminals on the back of the ammeter. Fit another wire from the second terminal to the terminal block. The circuit illustrated shows the position from each PSU where you can break into the supply and fit an ammeter.

You could also use your multimeter, or you could purchase a clamp ammeter: all will do the job. The clamp ammeter is the only one that does not require any rewiring.

BUILD YOUR OWN PSU

You may wish to build your own power unit and speed controller from scratch. The circuit in Fig. 3.14 shows a typical power supply circuit where the track speed controller does not have a transformer included.

The transformer would be a 12 volt AC output at 4 amps. The circuit breaker should be rated initially

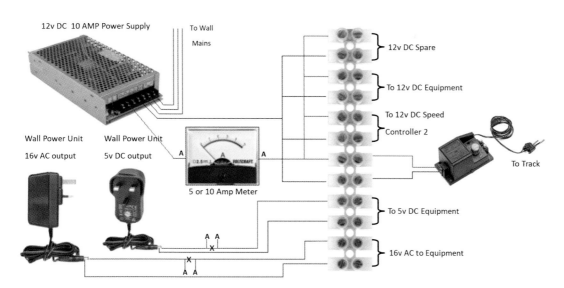

Fig. 3.13   *How to add an amp meter if you want to monitor the current being used.*

50 ELECTRICAL SYSTEMS AND POWER SUPPLIES

at 3 amps to protect your circuits and equipment from short circuits. The rectifier converts AC to DC and must be capable of carrying at least what the transformer is delivering, so 4 amps or above.

The speed controller can be any transformer-less unit from any proprietary manufacturer. Note that all wiring should be no less than 16/0.2 equipment wire.

Also note that the DC power supply below is unregulated and not stabilized, and will work well for most railway systems, and will run ancillary equipment with no problem. However, it is not recommended for use on sophisticated electronic circuits; the circuit described below will do this job. The circuit in Fig. 3.15 is primarily the same as the above, except for the transformer and rectifier; these have been replaced with the power supply that is shown in Chapter 2, Electronic and Electrical Components Explained.

We now have all the different types of power for our layout. Apart from the plug-in units this can all be put into a well ventilated enclosure ready for your layout.

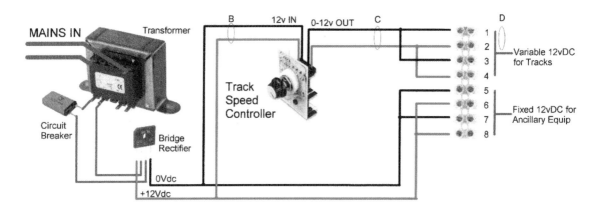

Fig. 3.14 *Example of a basic unregulated power supply using just three components.*

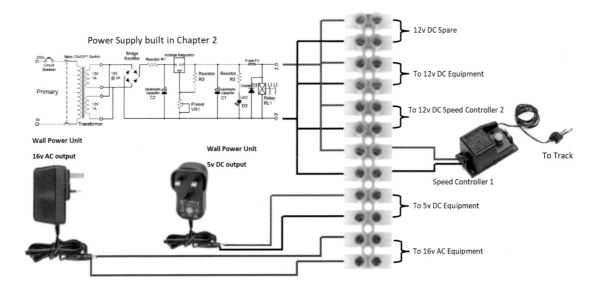

Fig. 3.15 *Circuit diagram showing a complete layout using the power supply built in Chapter 2.*

CONNECTING TO THE HOUSEHOLD MAINS

All this needs to be connected to a domestic socket wherever your layout is. You should bear in mind the following points.

First, start with an RCD – a residual current device – in the domestic socket. This is a device that instantly breaks an electric circuit to prevent serious harm from an ongoing electric shock. An RCD is designed to disconnect a circuit quickly and automatically when it detects that the electric current is not balanced between the energized (line) conductor(s) and the return (neutral) conductor. Under normal circumstances, these two wires are expected to carry matching currents, and any difference can indicate that a short circuit or other electrical anomaly is present, such as leakage. Leakage can indicate a shock hazard (or shock in progress), which is a potential danger to a person.

Second, use a multiway connector. There are many types and styles on the market, and some even come with built-in RCDs. Keep the cable from the household socket – RCD – to the multiway connector as short as possible. Do not use a coiled extension lead, or if you do, uncoil the lead before you use it. Plug all your power units into the multiway, and keep the wires to the control station as short as possible.

WHY UNCOIL EXTENSION CABLES?

A powered-up extension cable is the making of an electromagnet, albeit very crude; however, the coil can heat up, and this could then cause a fire. I must stress this is very unlikely with the currents you will be using, but it is always better to be safe than sorry.

CHAPTER FOUR

WIRE AND CABLES

In this chapter we are going to try and take the mystery out of the wiring and control of model rail systems. We are assuming that the reader has no knowledge of electricity or electronics, and have tried to set out successive points in the sequence they will present themselves.

In a model rail system your individual pieces of equipment may all use the same voltage, but will probably all have different current requirements. This is why we concentrate on the current rating of the wire, and not so much the voltage; most wire is rated higher than you will ever use. When you run two or more pieces of equipment you must add together the current used by each piece, and then check that the wire you are using can carry that amount of current. In practice, unless your layout is very large, you won't need to measure the current rating of each piece of equipment, as long as you keep to the following rules.

Equipment wire comes in two forms: solid or stranded. You will also need a basic multimeter: it should be able to read AC voltage just above 240v, DC voltage just above 40v, and resistance and current up to 10 amps. (See Chapter 11, Testing and Troubleshooting, to find out how to use a multimeter.)

PROTECTION

As most of the wiring is under the baseboard it is advisable to wear protective glasses when working under the board because offcuts of wire and even solder could fall on you and cause damage to your eyes and face. A good light source is also a great help.

SINGLE AND MULTI-STRANDED WIRE

Solid wire is composed of a single piece of metal wire, also known as a strand (conductor). Stranded wire is composed of many pieces of solid wire all bundled into one group. It is much more flexible than solid wire of equal size. Conversely, solid wire is used when little or no movement is needed. An example of this is your house wiring, which is all solid conductor because once it has been installed it is not going to move; most items that you plug into a socket, on the other hand, such as your iron or toaster, will have stranded conductors. For our purposes in general the more strands the wire has, the higher the current rating, and the larger the solid conductor, the higher the current rating.

The method of describing wire at first appears complicated, but in fact it is very simple once you know. The first number refers to the number of strands, and the second number refers to the

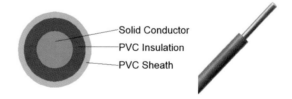

Fig. 4.1   Solid conductor.

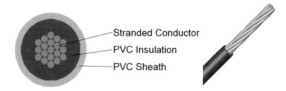

Fig. 4.2   Stranded conductor.

Wire gauge and where to use it

Catalogue size	Description	Current rating	Outside diameter	Where it can be used
1/0.25mm	1 strand of 0.25mm-diameter wire	0.5A	0.55mm	This is a high temperature resistant wire, and very small at half a millimetre in diameter. It is therefore ideal for wiring inside locos, carriages, signals, and any tight spaces
1/0.6mm	1 strand of 0.6mm-diameter wire	1.8A	1.2mm	A solid core wire. Not great on movement as it will break eventually, but ideal in those areas where once installed it is not going to move
7/0.2mm	7 strands of 0.2mm-diameter wire	1.4A	1.2mm	Use this for individual equipment back to a terminal block or distribution block
16/0.2mm	16 strands of 0.2mm-diameter wire	3.0A	1.6mm	This is a better wire to use from equipment back to the control panel or distribution boards: better to be higher than too low
24/0.2mm	24 strands of 0.2mm-diameter wire	4.5A	2.4mm	Ideal for bus bars on small layouts
32/0.2mm	32 strands of 0.2mm-diameter wire	8.5A	3.8mm	Ideal for bus bars on larger layouts

diameter of each strand in millimetres. For example, 16/0.2mm indicates sixteen strands of 0.2mm wire making one wire, and 1/0.6mm indicates one strand of 0.6mm wire making one wire.

All the above wires come in a range of colours: red, black, green, blue, yellow, grey, orange, violet, brown, white.

Stranded wire such as 7/0.2mm through to 32/0.2mm needs a little care when terminating it. Strip back the insulation and then twist the strands together as tightly as you can; soldering the stripped end is ideal. You must make sure that all the strands are in the terminal: just one 0.2mm wire left out – and one is hard to see – is enough to create a short with another wire. This is more critical the bigger the wire and the higher the current passing through it: remember your physics lessons at school where you were shown how electricity can jump from one conductor to another without the two touching each other.

1/0.6 equipment wire is by far the easiest to use because of its solid core; however, the drawback is that it is not very flexible, so it should not be used if the wire is going to move much.

When the current drawn exceeds the wire current rating it will start to heat up and eventually fail.

You will notice that the majority of wired equipment that you purchase has wires around the 1.2mm diameter, so is probably 7/0.2mm equipment wire. This is because the equipment does not individually draw more than 1.4 amps. However, when two or more pieces are run together, then the combined current draw could be more than 1.4 amps, so a larger feeder wire is required.

Running each wire back to the control panel is one way of reducing this over-current problem; however, you will eventually end up with hundreds of wires going back to the panel. If this is the way you want to go there is a range of multicore cables available, from 4-core to 25-core; each core has a unique colour code.

MULTICORE MULTI-STRANDED CABLES

Multicore cables are made up of a number of single core wires, each containing a certain number of strands put together in a plastic sheath. The most common cable you will have used is the power lead to your toaster or kettle. In a model railway system the first one you use is the power supply cable to the mains socket. These cables will be made

WIRE AND CABLES

Fig. 4.3 A range of multicore cables.

SCREENED MULTICORE CABLES

So why would you want to use screened wire?

All electrical wire and equipment creates noise, either radiated or conducted as EMI (electromagnetic interference), and this can seriously disrupt the proper operation of equipment near the wire. Here is not the place to investigate the science behind this, rather how we can safeguard against it.

It is important to state that the average model rail layout will not suffer from the effects of EMI, but if you are using sensitive electronic modules, as used in DCC systems, there is a greater possibility of this happening. The effects of EMI are that things happen that you don't expect, or that you have not initiated, or things don't work when they should.

up of either 3 × 16/0.2, 24/0.2 or 32/0.2 wire as seen above, depending on the current rating of the appliance they are powering. We will go into this in detail later.

The other multicore cables you will use are known as data or low power cables. These cables come in various sizes from 2-core to 25-core, and various strand sizes from 1/0.6 (this used to be called 'telephone' wire), 7/0.2 and 16/0.2.

Multicore Cable Description

Number of cores	Core size	Current rating	Outside diameter	Description
2-core power	16/0.2mm	2.5 amp	4.5mm	Used to power equipment that does not require an earth cable
	24/0.2mm	5 amp	6.10mm	
3-core power	16/0.2mm	2.5 amp	5.8mm	Used to power equipment that does require an earth cable. The 16/0.2mm can be used for bus bar feeds
	24/0.2mm	5 amp	6.10mm	
4-core power	16/0.2mm	2.5 amp	6.30mm	Used to power equipment that is three phase (we don't need to know about this). The 16/0.2mm and the 24/0.2mm can be used for bus bar feeds
	24/0.2mm	6 amp	7.70mm	
4-core data	7/0.2mm	1.4 amp	3.85mm	Used to send small power signals to equipment around the layout. The 16/0.2mm will handle currents up to 2.5 amp
	16/0.2mm	2.5 amp	6.3mm	
6-core data	7/0.2mm	1.4 amp	4.55mm	
	16/0.2mm	2.5 amp	7.3mm	
8-core data	7/0.2mm	1.4 amp	5.60mm	
	16/0.2mm	2.5 amp	8.8mm	
12-core data	7/0.2mm	1.4 amp	6.10mm	
18-core data	7/0.2mm	1.4 amp	7.40mm	
25-core data	7/0.2mm	1.4 amp	8.70mm	

WIRE AND CABLES

Fig. 4.4 Screened multicore cables.

So how do we solve this?

We must mention at this point that some people say that twisting the cable together solves the problem; there is some truth in this, but it can be very difficult to achieve. In straight cable, all noise current is flowing in the same direction. When the cable is twisted, in some parts of the signal lines the direction of the noise current is the opposite from the current in other parts of the cable. Because of this, the resulting noise current is many factors lower than with an ordinary straight cable.

The alternative is to use screened cable. The shield can act on the EMI in two ways: first, it can reflect the energy, and second, it can pick up the noise and conduct it to ground. (This is important: if you use screened cable you must earth one end, otherwise it has nowhere to put the noise.)

Fig. 4.5 Twisted wires.

Fig. 4.6 Make-up of screened cable.

Normally, screened cable has a foil sleeve, then wire braid, and finally the plastic outer protective covering. The braid is a woven mesh of bare or tinned copper wires, which provides a low-resistance path to earth. The effectiveness of the braid depends on the tightness of the weave: EMI will get through the holes in the weave, so the tighter the weave, the better.

A shield system is only as good as its weakest component. If you don't tie the braid to earth, then it won't work. It is therefore a waste of time and money to place shielded cable between points A and B without earthing it.

SOME BASIC PRINCIPLES

- The biggest problem with layouts is that they tend to get bigger and more complex as the months go on, with this in mind fit slightly larger cables from day one. This means you will not have to keep changing the wire size as the load increases
- Most of the points and signals you have installed have three wires – a negative (0v) and two positive wires – and the tracks have two wires, one negative and one positive. We are unable to give you colour codes here, as all manufacturers appear to use different colours for certain functions. As you will soon be aware, the positive wires are the ones you use to control the equipment via speed controllers or switches
- Colour code your wiring from the start of your project, allocating colours to specific tasks: red for uncontrolled (not switched) positive voltage, blue for controlled (switched) positive voltage, black for negative voltage, your main feed and return. Points, green and orange; signals, white and yellow – and so on
- Keep the wires tidy with cable ties and sticky pads. If you use marker tags you can identify the cables as you go. Keep a cable and wire register, which can be downloaded and printed. There is nothing worse than trying to find a faulty wire from a bird's nest of wires

TERMINATION OF WIRES

The equipment you have purchased will normally have short wires (100 to 200mm long), which will need to be terminated in a terminal block under the baseboard. From this terminal block extension wires are run to the control panel, and these wires must be selected correctly. If they are just powering the one piece, then the extension wire can be run in 7/0.2 or 16/0.2; however, if they are common wires they may need to be larger – 24/0.2 or 32/0.2 – or a bus bar can be used.

There are many types of terminal block, but essentially it comes down to a personal choice: do you want the screw type or the solder type? Here we show just a few. The basic terminal block (some people call it a 'chocolate block') normally comes in a length of twelve terminals and various current ratings, but it can be cut to any length required.

The pluggable terminal block again comes in twelve terminals and various current ratings, and is easy to cut into desired lengths. These blocks are ideal for places where you have removable sections of layout, or your whole layout is demountable.

The three-way termination block is easily fitted near points and signals to take the wires from the equipment, then extend them to wherever you need them. They are supplied in a six-way unit, but each three-way section can be snapped off.

Common voltage junction blocks are solid brass and are available in four- or eight-way sections. They are ideal for terminating all your negatives, without having to make loops or overloading a single terminal with too many wires.

Wire clips are very useful for adding a spur to an existing wire without cutting the existing wire. Place the existing wire in the back groove, then place the new wire in the front groove, and press the two halves together. There is no need to strip the wire. They come in two sizes.

Fig. 4.7 Twelve-way terminal block (sometimes called a chocolate block). JPR ELECTRONICS

Fig. 4.8 Twelve-way pluggable terminal block. JPR ELECTRONICS

Fig. 4.9 Three-way in/out terminal block. BRIMAL

Fig. 4.10 A brass common terminal block. CPC

Fig. 4.11 Junction clip for adding extra wire without cutting the primary wire. JPR ELECTRONICS

WIRE AND CABLES 57

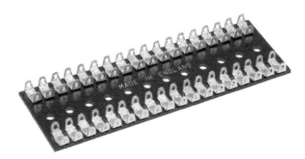

Fig. 4.12 Thirty-six-way solder junction board.
RS COMPONENTS

If you prefer to solder your wires, there are products available to do this. The most common solder terminal strip is shown in Fig. 4.12. Here you get thirty-six individual double tag points to solder to.

KEEPING WIRES AND CABLES TIDY

It is essential that you keep all cables and wires as neat as possible from day one. Decide on four or five main routes around your layout, and remember – if anything goes wrong you will be working upside down. There are a number of products that can help with this.

The cable clip: This is the best known cable clamp. They are ideal for multicore cables, but not so good for individual wires. They come in a range of sizes, usually with a nail to fix them in place. Nailing things to the underside of a baseboard is not a good idea, though it is all right to do so on struts or legs.

The cable tie and base: These are probably the best to use, as you can fit them around the baseboard and leave them full open until you have finished. Just thread the wires through until you need to terminate them.

Aluminium clip: These are a great idea but they will not hold the cables while you are assembling the layout. They need to be closed as you put the cable or wire in place, and they can be opened and closed a few times, but being aluminium they will eventually break.

The P clip: This is another great idea, as long as you pick a large diameter hole from the start – then you can add cables and wires as you expand.

Fig. 4.14 Cable tie and base. Cable tie is available in many different lengths. JPR ELECTRONICS

Fig. 4.13 Standard cable clip available in many sizes.

Fig. 4.15 Aluminium cable clamp with self-adhesive base. RAPID ELECTRONICS

WIRE AND CABLES

Fig. 4.16 The P clip, which is available in many diameters. JPR ELECTRONICS

Fig. 4.17 Cable trunking, available in many widths and depths. RAPID ELECTRONICS

Please note the problem with EMI, as mentioned above. Try to keep pure power cables away from instruction cables, as low power instruction cables (signal cables) can be susceptible to EMI.

CABLE MARKING

From day one, try to devise a system for marking cables. As your layout grows you will find it harder to identify wires when they are in a large bundle. First decide on the use of the nine basic colours, such as red for uncontrolled (not switched) positive voltage, blue for controlled (switched) positive voltage, black for negative voltage, your main feed and return; points, green and orange; signals, white and yellow; and so on.

There are a few products that can help with wire identification. You need to devise a table to log each wire so you know where it comes from and where it is going. I devised the table shown below.

Cable-tie markers: These are useful for a bunch of cables if all are going to a specific area.

Miniature cable trunking: This is also very useful but it has one disadvantage: as you are working upside down, each time you take the cover off to fit a new cable all the existing wire and cables are going to fall out. It is best used on legs, or areas you know you will not be adding to.

You will find that as your layout progresses a combination of all of the above will be included on your layout.

Fig. 4.18 Cable ties with a write-on marker plate. RAPID ELECTRONICS

Control panel wiring chart

Code no.	From	To	Size	Colour	Description
SW12C1	Control panel 12	Points no. 12	16/0.2mm	Blue	Up wire for points 12
SW12C2	Control panel 12	Points no. 12	16/0.2mm	Yellow	Down wire for points 12
SN10C1	Signal switch 10	Signal 10	7/0.2mm	Brown	Green for signal 10
SN10C2	Signal switch 10	Signal 10	7/0.2mm	Orange	Red for signal 10

WIRE AND CABLES

Fig. 4.19 Clip-over cable identification bands.
RAPID ELECTRONICS

Cable clip-on markers: These come in rolls or dispensers, and are available in all numbers and the complete alphabet. I use these to create the codes above for each wire or for a bunch of wires. It is possible to fit three to four 7/0.2mm wires on each marker.

CHAPTER FIVE

CONTROL PANELS

The control panel is the single most important part of your layout, and is something you have to make yourself, as it will be unique to your layout. The control panel will be where the speed controllers are located, and all the switches for the points, signals, and any other ancillary equipment you have on your layout. Some people will go for the option of placing all the controls in cut-outs on your baseboard, others will go for mounting switches and so on in pre-made desktop enclosures. Both systems are fine; however, there are a few things you need to remember, so we will start with the baseboard cut-out method.

INSET CONTROLS

A number of manufacturers have panel-mounted speed controllers, but the majority have desktop control units. If you opt for a panel-mounted controller it will not come with a transformer, so you will need space under the controller for the transformer and a terminal block. Use the manufacturer's instructions to determine the size of the hole to cut.

Fig. 5.1 Gaugemaster dual-speed controller.
GAUGEMASTER

Fig. 5.2 Vesta dual-speed controller. VESTA

Figs 5.1 and 5.2 show examples of insert-type control panels; although the second one comes in an enclosure, it can be removed and set into a control panel. In both cases the transformer needs to be mounted underneath the controller, and then wiring to and from the controller. The basic circuit is shown in Fig. 5.3, but use the manufacturer's instructions for the final details.

The wiring from the transformer to the controller and then to the tracks should be run in a minimum of 16/0.2mm wire.

Ensure the transformer is well ventilated, and at the same time protected from accidental touching. The wires on to the transformer should be soldered in place, with each terminal protected with a short length of heat-shrink tubing to protect the bare metal. The mains input to the transformer should go through a fuse, with a current rating of 2 amps. The earth wire (yellow/green) from the plug should be bolted to the chassis frame of the transformer.

Each manufacturer will have their own method of termination to the speed controller, so read their instructions in conjunction with the above.

The next part of the control panel is the switch control area. Here you will have all the switches required to activate your points, signals, building

CONTROL PANELS

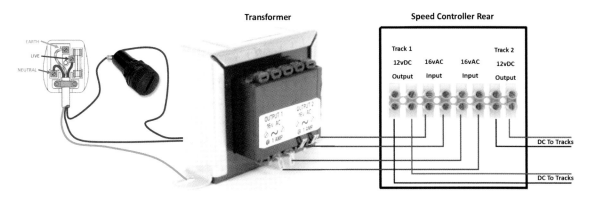

Fig. 5.3 Connecting some speed controllers to power will require fitting an appropriate transformer.
BRIMAL

lighting, and anything else you may require. As an example, let's use the layout in Fig. 1.24 of Chapter 1, The Layout.

The switches and so on can be mounted on any material up to 5mm thick, as most switches and indicators have that amount of thread available. I personally would opt for 3mm thick, clear Perspex. If you limit it to A4 landscape you can *reverse* print on to a clear label all the names and codes, even your layout, and drill points. I say 'A4' as most people will have an A4 printer – anything bigger and you will need to go to the expense of an A3 printer.

This is then stuck to the Perspex, taking care not to get any bubbles: a rolling pin or wallpaper edge roller will give a bubble-free finish. Cut any excess label from the edges now, and fit Sellotape to the label about 5mm in and then up the side of the Perspex, and trim. This is done to stop the label from coming away from the Perspex while you are working on it. The label now needs to be overpainted to protect the printing. The more coats the better, as the painted surface is going to be face down while you drill all the holes.

Figs 5.5 to 5.10 are some of the most common items that will be placed on the control consol.

For this control panel we may not use all these items. The layout ready to print on to a clear label in *reverse* could look like that shown in Fig. 5.11.

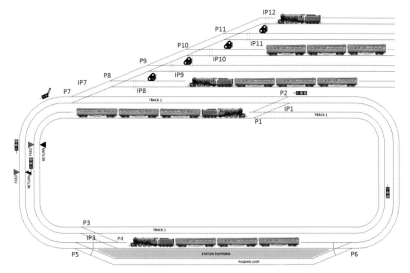

Fig. 5.4 The complete layout that you are going to transfer to your mimic control panel.

CONTROL PANELS

Items that will be placed on the control consol

3mm LED holder		5mm LED holder	
	Hole diameter: 4.5mm		Hole diameter: 6.5mm
	Bezel diameter: 5.0mm		Bezel diameter: 8.0mm
	Max. panel thickness: 3mm		Max. panel thickness: 6mm
Fig. 5.5 3mm LED holder. JPR ELECTRONICS		**Fig. 5.8 5mm LED holder.** JPR ELECTRONICS	
Standard push switch		Miniature push switch	
	Hole diameter: 11.65 (12mm)		Hole diameter: 7.5mm
	Bezel size: 15.5mm square		Nut bezel diameter: 11.0mm
	Max. panel thickness: 7.0mm		Max. panel thickness:
Fig. 5.6 Standard square push-button switch. JPR ELECTRONICS		**Fig. 5.9 Miniature small head push-button switch.** JPR ELECTRONICS	
Miniature toggle switch		Rotary switch	
	Hole diameter: 6.5mm		Hole diameter: 9.5mm
	Nut bezel diameter: 9.5mm		Nut bezel diameter: 15mm
	Max. panel thickness: 6.0mm		Max. panel thickness: 7.0mm
Fig. 5.7 Miniature toggle switch. JPR ELECTRONICS		**Fig. 5.10 Rotary switch.** JPR ELECTRONICS	

Once you have decided on your control panel layout, you should next design the layout for all the terminations underneath the panel. Again, a good way to do this is on A4 (300 × 210) size, as you can print A4 labels and then stick these on the mounting board before you fix the terminal blocks, so that you have all the labelling done. Obviously you may have more than one A4 board. The termination layout for the above control panel could look something like that shown in Fig. 5.12.

To power up the switches in your control panel, use one of the power supplies mentioned in Chapter 3, Electrical Systems and Power Supplies. Arriving into the control panel area you will want 12v DC, 16v AC, and possibly 5v DC, as shown in the top left-hand corner of Fig. 5.12. Then you have a common negative, and common positive 12v DC bus bars. These are made from solid chromed brass and are readily available. Below this you have two terminals for each switch, all labelled for easy identification.

In the bottom left-hand corner is the terminal block to and from the speed controllers, depending on how many you have. You will need a second mounting board for the timers and lighting controllers, which will fit to the right of this one.

It is important to have easy access to this termination board, so rather than having to work upside down, if you make the switch panel easily removable then access can be from the top.

CONTROL PANELS

Fig. 5.11 Example of switch and indicator control panel based on the layout in Fig. 5.4.

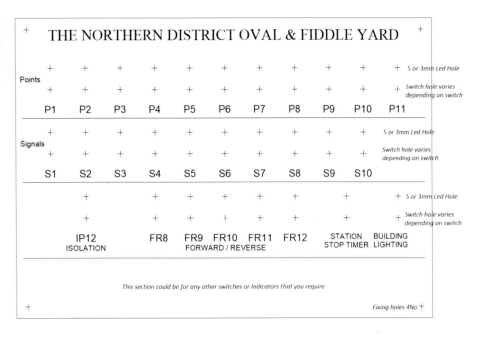

Fig. 5.12 Termination board under the control panel based on the layout in Fig. 5.4.

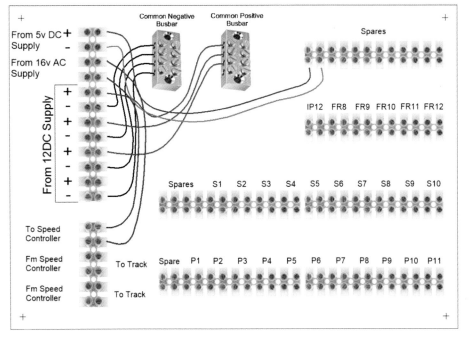

In Fig. 5.13, each points switch is going to require at least three wires. The common positive that goes to the centre terminal on the switch will come from your 'common positive bus bar'. The other two wires from the switch go to the point motor via the terminal block S1. The negative for the same point motor will be supplied from the 'common negative bus bar'.

For full wiring of all equipment and individual functioning switches, *see* Chapter 7, Track Wiring.

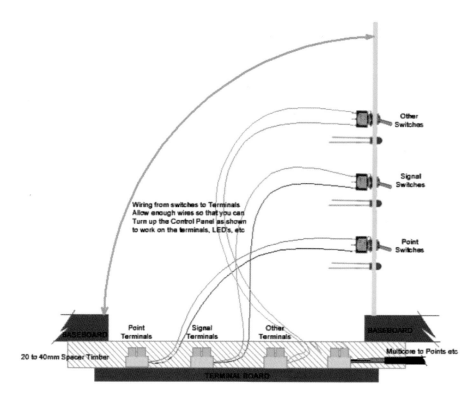

Fig. 5.13 Fitting the control panel into your baseboard or separate control enclosure. BRIMAL

DESKTOP CONTROLS

If you prefer a desktop-type console, there are plenty about: some are plastic enclosures with aluminium front plates, others are aluminium top and bottom. The basic principle is the same. The internal termination board can be either a removable plate, or everything can be fixed to the enclosure's base. You will need a lot of holes on the back for all the cables going to the layout.

The front panel is slightly different from the one described above, as the panel is aluminium, or painted steel. The front panel can be printed on to a metallic (or any type) of label, and then stuck on to the enclosure's front plate. If you have a border (as the first two examples show) then the label needs to be a little bigger so that the edges can be tucked in and stuck to the back of the panel. There is nothing worse than the label starting to unstick in the corners. The other two examples have only left and right borders, so the top and bottom of the label will have to rely on its stickiness to keep it in

Fig. 5.14 Sloped desktop control panel (faceplate 215 × 130mm). JPR ELECTRONICS

Fig. 5.15 Sloped desktop control panel (faceplate 170 × 143mm). RAPID ELECTRONICS

CONTROL PANELS 65

Fig. 5.16 Sloped desktop control panel (faceplate 215 × 130mm).

Fig. 5.17 Same size as Fig. 5.16 but with a higher back to allow for transformers internally.

place. Another way is to use a thin (1 or 2mm) sheet of Perspex to cover the label, and position it so it is held in place by the switches.

However you decide to manufacture your control panel, the most important part of the panel is having the correct switch for the function you want.

CHAPTER SIX

SWITCHES

In a model railway layout you will require a number of different types of switch, each with its own specific function. This chapter provides a description of the switches that you will require and their function.

There is only one switch that is unique to model railway systems, and this is the passing contact switch, which was designed to look like the real thing in a signal box. Mechanically it works well, however electrically it does not. To make the switch similar to the real thing and also only supply a small pulse to the motor, the lever has to pass over a contact and stay in the position past the contact to show the position of the points.

Switches come in many shapes and sizes, and have many functions, and there are many ways of actuating a switch, from push button to magnetism. They all have many specifications regarding their make-up, but there are two that you should always check before you purchase a switch: the switching current, and the making and breaking capacity.

The switching current: Most switches will be able to handle the voltages you are going to be using, but will they handle the current? Examples can vary from a standard toggle switch – Fig. 6.24 can carry 10A per contact – whereas the reed switch can only handle 100mA (Fig. 6.40).

The making and breaking capacity: When the two contacts of a switch join, there is a point where they have not joined but the electricity has started to flow by arcing between the contact plates. This can happen when switching 'on' and also when switching 'off'. This is good and bad at the same time. The good part is that it cleans the contacts of dirt to ensure a good contact, the bad part is that if it is too high it starts to burn the contact surface and eventually the contacts will not join properly. Some switches will have a minimum switching current and a maximum switching current, and a current carrying capability specified. Keep an eye on the 'maximum switching current' if specified. As a rule of thumb if you are not switching near the capacity of the switch then you will be all right.

An example of this is, don't try and switch a point motor with a reed switch, as something will go bang eventually. Use the reed switch to power a relay that has the current capability to switch the point motor on and off.

THE PASSING CONTACT SWITCH

The first thing that you need to know is that point motors can be either motors or solenoids. The passing contact switch is designed to work solenoids, which only require a short pulse of electricity to energize the solenoid. This is why proprietary manufacturers call them 'passing contact switches', because as you change the position of the lever you pass a contact, which energizes the motor for a fraction of a second. The problem occurs when you want to change the points again, and the switch has to pass the first contact again before it gets to the next contact to change the points. This causes the points to 'chatter', or 'bounce'. The momentary toggle switch has become very popular for this reason, as it does not pass an already made contact to get to the other contact. The down side is that it doesn't look like the real thing, it looks like what it is – a toggle switch; also it is always in the centre, and therefore does not show the position of the points.

The illustrations give a few examples of passing contact switches.

SWITCHES

HOW DOES A PASSING CONTACT SWITCH WORK?

For this explanation let's say 'A' represents the points straight through, and 'B' the 'turnout'. The position of the lever indicates that the switch has just put the points in the straight-through position; the lever is in the 'up' position, which tells you where the points are. Now to change the points the lever has to pass 'A' to get to 'B', thereby energizing the motor again, which makes it 'chatter' (because the solenoid is already there) before it reaches 'B' to energize the motor in the other direction.

All passing contact switches have the same circuit, just different ways of achieving it.

As can be seen by the circuit diagram in position 1, the actuating lever is to the right of A1, which means it has energized the A1 coil. To energize the

Fig. 6.1 Eckon EE4 passing contact switch. ECKON

Fig. 6.2 Hornby R033 passing contact switch. HORNBY

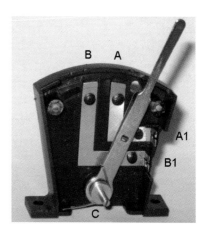

Fig. 6.4 Internal image of a passing contact switch.

Fig. 6.3 Peco PL-26 passing contact switch. PECO

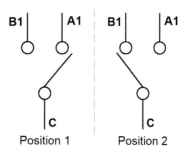

Fig. 6.5 Schematic diagram of a passing contact switch in both positions.

B1 coil it has to pass A1 (energizing it again), and then energize B1. The circuit diagram is different from all conventional switches because of this passing contact ability.

Other types of switch that can be used for solenoid-type point control can be seen in the following illustrations.

Fig. 6.6 Miniature momentary toggle switch.
JPR ELECTRONICS

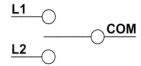

Fig. 6.6a Schematic diagram of momentary toggle switch.

Fig. 6.7 Standard momentary push-button switch (you will require two for each point set).
JPR ELECTRONICS

Fig. 6.7a Schematic diagram of a momentary push-button switch. BRIMAL

Fig. 6.8 Miniature momentary push-button switch (you will require two for each point set).
JPR ELECTRONICS

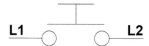

Fig. 6.8a Schematic diagram of a miniature momentary push-button switch.

All the switches above must be normally open momentary action switches. The first example is a mini toggle switch biased to the centre with momentary 'on' in both left and right directions. With this switch you can control both solenoids with the one switch. The next two types will require two switches for each point set, one for each solenoid coil.

On the point motor there will be three wires from it: one is the negative supply, and the other two go to coil 1 and coil 2. Manufacturers use different colours for these wires, so you will need to see the instruction sheet that comes with your point motor.

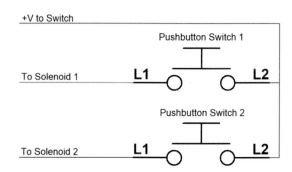

Fig. 6.9 Circuit diagram for using two push switches to control a point set. BRIMAL

SWITCHES

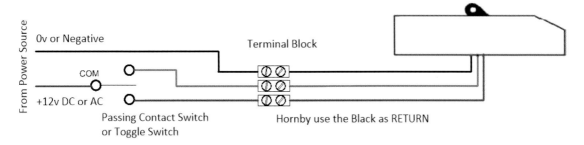

Fig. 6.10 *Wiring diagram of a mini toggle switch to a point motor.*

This is an example of the wiring of most solenoid-type point motors.

THE ON–OFF SWITCH

This basic on–off switch has many uses on your layout: for example, controlling isolated sections of track or to control street lights, building lighting, and many accessories that just require a simple on–off switch. All proprietary manufacturers make an on–off switch to complement their range; a few examples are illustrated here.

Fig. 6.11 *Hornby R047 on/off switch.* HORNBY

Fig. 6.12 *Peco PL-22 on/off switch.* PECO

Fig. 6.13 *Cobalt on/off switch.* DCC CONCEPTS

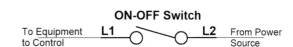

Fig. 6.14 *Circuit diagram for an on/off switch.*

All switches have the same circuit, just different ways of achieving it.

Other types of switch that can be used for on–off functions are shown in the following illustrations.

The push-button switches must have a latching function, as opposed to momentary. The slide switch is normally available in on–on configuration; it is very rare to find a two-terminal slide switch, in which case you would use the centre terminal and one other to create an on–off switch. The toggle switch will need to be latching: in other words, it stays in the last state until it is operated again; if the toggle switch

Fig. 6.15 Standard latching push-button switch.
JPR ELECTRONICS

Fig. 6.15a Schematic diagram for a standard latching switch.

Fig. 6.16 Miniature latching push-button switch.
JPR ELECTRONICS

Fig. 6.16a Schematic diagram for a miniature latching push-button switch.

Fig. 6.17 Standard size slide switch.
RAPID ELECTRONICS

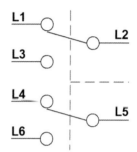

Fig. 6.17a Schematic diagram for a slide switch.

Fig. 6.18 Miniature toggle switch.
JPR ELECTRONICS

Fig. 6.18a Schematic diagram of a miniature toggle switch.

has brackets around the 'on' then it is momentary and cannot be used in this situation.

THE ON–ON SWITCH

These switches are used to divert power from one circuit to another. Two-aspect coloured light signals switch from red to green and back again; they can also switch power from one section of track to another, and are designed to divert power from one circuit to another. This switch is also used to operate certain turntables.

All switches have the same circuit, just different ways of achieving it.

Other types of switch that can be used for on–on functions are shown in the following illustrations.

SWITCHES 71

Fig. 6.19 Hornby R046 on/on switch. HORNBY

Fig. 6.20 Peco PL-26R on/on switch. PECO

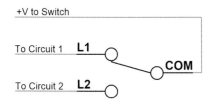

Fig. 6.21 Circuit diagram of an on/on switch.

Fig. 6.22 Miniature on/on toggle switch.
JPR ELECTRONICS

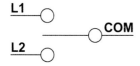

Fig. 6.22a Schematic diagram of the miniature toggle switch. BRIMAL

Fig. 6.23 Standard slide switch DPDT.
RAPID ELECTRONICS

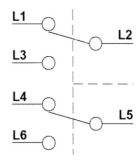

Fig. 6.23a Schematic diagram of a DPDT slide switch. BRIMAL

Fig. 6.24 Standard toggle switch SPST.
JPR ELECTRONICS

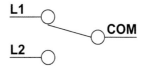

Fig. 6.24a Schematic diagram of an SPST toggle switch.

Fig. 6.26 shows how an on–on switch controls a two-aspect signal: the green light is on – changing the switch will change the green to red.

Fig. 6.27 shows how an on–on switch is used to control sections of track. The tracks are connected with plastic 'fishplates', which isolate the left track

SWITCHES

Fig. 6.25 Rotary switch four-pole three-way.
JPR ELECTRONICS

from the right. The switch at the moment shows that the right track is powered and the left track is isolated. By changing the switch the live track can be changed to the left track, isolating the right track.

THE ON–ON–ON SWITCH

This switch has a number of functions. It can be used to control three-aspect signals, as you have three positions, each position controlling one of the three signal lights. It can also be used for controlling power to sidings in a fiddle yard.

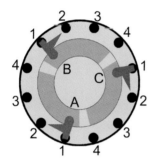

Fig. 6.25a Schematic diagram of a rotary switch.

Fig. 6.28 Miniature toggle switch on/on/on DPDT.
JPR ELECTRONICS

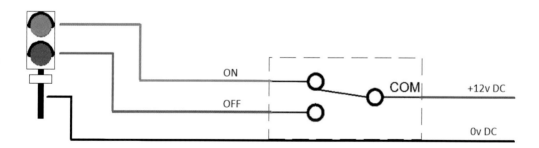

Fig. 6.26 A simple circuit using an on/on switch to control a two-aspect signal. BRIMAL

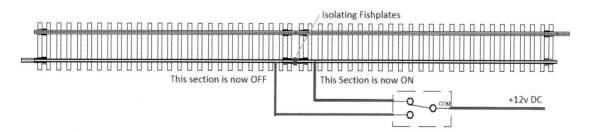

Fig. 6.27 A simple circuit using the same on/on switch to isolate a section of track. BRIMAL

SWITCHES

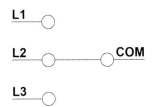

Fig. 6.28a Schematic diagram of the three-way on switch. BRIMAL

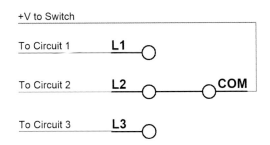

Fig. 6.29 Circuit showing the switch used to control three separate circuits.

Fig. 6.30 shows how the switch is used to control a three-aspect signal. The position of the switch arm shows which LED is on in each position.

THE CENTRE OFF SWITCH ON–OFF–ON DOUBLE POLE (DPDT)

This configuration is the same as the passing contact switch, except this latches in each position. This switch is used to reverse polarity at the main control panel or a section of track. It can also be used to switch power from one piece of equipment to another. In most cases these switches are double or triple pole. It is always better to switch both the positive supply and the negative supply rather than just one, hence the need for a minimum of a double pole switch.

Fig. 6.31 Rotary switch four-pole three-way. JPR ELECTRONICS

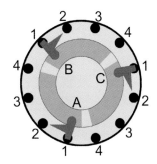

Fig. 6.31a Schematic diagram of a rotary switch on/off/on.

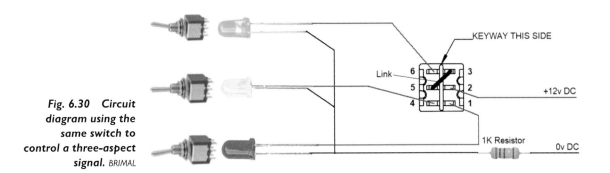

Fig. 6.30 Circuit diagram using the same switch to control a three-aspect signal. BRIMAL

74 SWITCHES

Fig. 6.32 Miniature toggle switch DPDT.
JPR ELECTRONICS

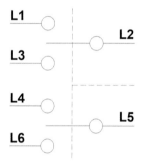

Fig. 6.32a Schematic diagram of the toggle switch DPDT.

Fig. 6.33 Standard slide switch DPDT.
JPR ELECTRONICS

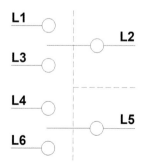

Fig. 6.33a Schematic diagram of the slide switch DPDT.

MICRO SWITCHES

Micro switches come in many sizes, and most are called changeover switches. They are also known as snap-action switches. They are notable for being able to change state with minimal physical force. There are a number of actuators available (the actuator is the part that actually throws the contacts from one state to another). The basic micro switch is used to sense a change of state on your layout: this could be a set of points, the position of a turntable, and so on.

The micro switch has the same switching configuration, which is a change-over contact. Therefore, one or the other contact is always made.

Fig. 6.34 Roller-type micro-switch changeover contacts. JPR ELECTRONICS

Fig. 6.35 Lever-type micro-switch changeover contacts. JPR ELECTRONICS

Fig. 6.36 Button-type micro-switch changeover contacts. JPR ELECTRONICS

SWITCHES

Fig. 6.37 Roller-type sub micro-switch changeover contacts. RAPID ELECTRONICS

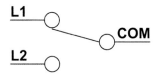

Fig. 6.38 Schematic diagram of changeover micro switches.

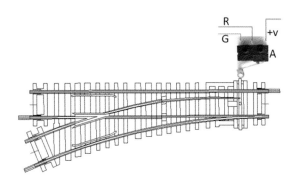

Fig. 6.39 Micro switch being used to indicate the position of a set of points.

Fig. 6.39 shows how a roller lever micro switch is wired to indicate the position of a set of points. The R and G can be wired to red or green indicators for a visual indication as to the state of the points.

THE MAGNETIC SWITCH

The magnetic switch is more commonly known as the reed switch. The reed switch consists of a pair of contacts in a hermetically sealed glass tube, which is operated by a magnetic field. The contacts can be normally open, normally closed, or changeover. The reed switch can be actuated by a coil or a magnet; once the influence of the magnetic field is removed, the switch will revert to its original state. The most common use that you will be aware of is in home security on doors and windows, with the switch on the fixed part and the magnet on the moving part. In a model railway situation, the magnet is on the underside of the train or the 'tie bar' of a set of points. The reed switch is placed between the tracks and is actuated when the train passes over it. As with most switches, there are many shapes and sizes: the ones selected here have been used on model railway layouts for many years.

Fig. 6.40 A range of different types of magnetic switch. JPR ELECTRONICS

Fig. 6.40a Normally open contact.

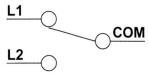

Fig. 6.40b Changeover contact. BRIMAL

Fig. 6.40c Normally closed contact. BRIMAL

Reed switches in glass form are very delicate and should be handled with care – the one illustrated is only 10mm long. See the section on installing reed switches for the best method of bending, cutting and soldering them.

THE SPRING-LEVER SWITCH

A spring-lever switch has a C spring to snap it from one state to the other. The switch has different actuation forces, dependent on the strength of the C spring: this can vary from 0.25N to 0.6N. Most spring-lever switches used on model railways have changeover contacts, as shown. They are not latching switches, so when the pressure is removed the switch will snap back to its resting state.

The drawing shows how a spring-lever switch is wired to indicate the position of a set of points. The R and G can be wired to red or green indicators for a visual indication as to the state of the points.

THE ROTARY SWITCH (Figs 6.45 to 6.45d)

A rotary switch consists of a spindle with a contact arm that projects from its surface like a cam. It has an array of terminals, arranged in a circle around the rotor, some close to the centre and some further away from the centre. The inner terminal is the main feed into the switch (shown in red in the diagram opposite); the outer terminal is the switched terminal output (shown as a black dot in the diagram opposite). Rotary switches use a 'star wheel' mechanism to provide the switching positions, such as at every 30, 45, 60 or 90 degrees. Nylon cams are then mounted behind this mechanism, and spring-loaded electrical contacts slide around these cams. The cams are notched or cut where the contact should close to complete an electrical circuit.

Fig. 6.41   Spring lever switch (leaf switch).
RS COMPONENTS

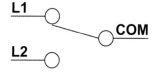

Fig. 6.42   Schematic diagram of a spring lever switch.

Fig. 6.44   An on/off rotary switch SPST also available in DPST. JPR ELECTRONICS

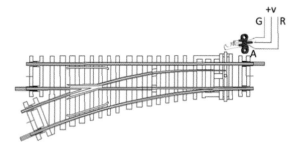

Fig. 6.43   A spring lever switch being used to indicate the position of a set of points.

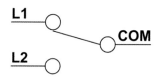

Fig. 6.44a   Schematic diagram of on/off rotary switch.

SWITCHES

Fig. 6.45 Multipole rotary switch.
JPR ELECTRONICS

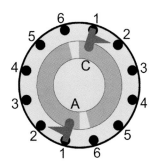

Fig. 6.45c Two-pole six-way rotary switch.

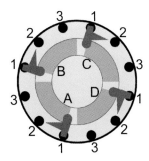

Fig. 6.45a Four-pole three-way rotary switch.

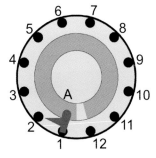

Fig. 6.45d One pole twelve-way rotary switch.

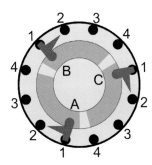

Fig. 6.45b Three-pole four-way rotary switch.

Some rotary switches are user-configurable in relation to the number of positions. A special toothed washer that sits below the holding nut can be positioned so that the tooth is inserted into one of a number of slots in a way that limits the number of positions available for selection. For example, if only four positions are required on a twelve-position switch, the washer can be positioned so that only four switching positions can be selected when in use.

The first illustration is of a high amp switch to be used for your main system on–off. The conventional rotary switches have a current carrying capability of 300mA per contact. The red mark shows the common terminal and the direction of travel.

TOGGLE SWITCHES

Miniature toggle switches have been mentioned many times above, as they can be used in most switching situations. The following table should show all the toggle switches that you might need to use on your layout.

Toggle switch connections

Description	Image	Contact arrangement	Termination end
Single pole, single throw (SPST) A basic on–off switch		L1 ○――○ L2	KEYWAY THIS SIDE; 2, 1
Single pole switches (SPDT) on–off, on–on, (on)–off; () denotes a momentary action.		L1 ○, L2 ○, COM	KEYWAY THIS SIDE; 3, 2, 1
Single pole switches (SPST) on–off–on, (on)–off–(on); () denotes a momentary action		L1 ○, L2 ○, COM	KEYWAY THIS SIDE; 3, 2, 1
Double pole, double throw (DPDT) on–off, on–on, (on)–off, on–off–on; () denotes a momentary action		L1 ○, L2, L3 ○, L4 ○, L5, L6 ○	KEYWAY THIS SIDE; 6, 5, 4, 3, 2, 1
Double pole, double throw (DPDT) on–off–on, (on)–off–(on), (on)–off–on; () denotes a momentary action		L1 ○, L2, L3 ○, L4 ○, L5, L6 ○	KEYWAY THIS SIDE; 6, 5, 4, 3, 2, 1

(Continued)

SWITCHES

Description	Image	Contact arrangement	Termination end
Triple pole, double throw (3PDT) on–off–on, on–on, on–off		L1, L2, L3, L4, L5, L6, L7, L8, L9	KEYWAY THIS SIDE 9 6 3 8 5 2 7 4 1
Four pole, double throw (4PDT) on–off, on–on, on–off–on		L1, L2, L3, L4, L5, L6, L7, L8, L9, L10, L11, L12	KEYWAY THIS SIDE 12 9 6 3 11 8 5 2 10 7 4 1

CHAPTER SEVEN

TRACK WIRING

BASIC PRINCIPLES

Model railway layouts tend to work on two systems: one is called 'analogue control', and the other, introduced in the 1990s, is the 'digital command control' (DCC). This section concentrates on the analogue control system. In an analogue system the speed and direction is all in the 12v DC placed on the tracks by the speed controller. The direction of the train is decided by which rail is positive and which is negative, and the speed is changed by adjusting the voltage put on the rail by the controller from 12v to 0v. We must note here that the variable voltage from 0V to 12v is not altogether true, because modern power supplies now use a principle called 'pulse width modulation'. However, for the purposes of this chapter the voltage varies from 0v to 12v to control the speed.

To supply this voltage to the rails you need to have a speed controller, which varies the output to the rails, thus enabling the locomotive to change speed. There will also be on the controller a forward–off–reverse switch. This switch can either remove power from the tracks, or reverse the polarity of the track, putting the train in reverse. The problem you will notice is that you can only run one train on the track because the track is controlling the train. To run more than one train – in fact several trains – we have to create isolated sections with a separate controller in the isolated section. Most points are called 'isolating points', so this helps.

To summarize, each train must have an isolated section of track with a separate controller. The positive rail and negative rail will change when the forward/reverse switch is used. This is confusing to show in diagrams, so instead of designating a rail +12v, we call it the 'feed' rail, and the negative rail becomes the 'return' rail.

WIRING COLOUR CODES

Before we get started we must set up a system of wire colour coding. This will help later with troubleshooting, and with rewiring the layout when you want to change something. The choice of colours is down to your personal preference, but it is important that you keep to it, because as your layout grows there will be a great many wires under the baseboard.

Allocate colours to a specific task: for example, red for uncontrolled (not switched) positive voltage, blue for controlled (switched) positive voltage, and black for negative voltage, your main feed and return. Points might be green and orange, signals white and yellow, and so on.

Note: be careful with purchased equipment that comes with short pieces of wire. Do not assume anything: check what each colour does. *You* may have used black for negative, but the black on a point set may not be the negative input to that point set. In the chart below the start colour is what is supplied with the equipment, and you may have to change this immediately.

Create a chart so that you can register the colours for each piece of equipment: an example of how to set it out is shown here. This will be populated as we

WIRE COLOUR CODING							
EQUIPMENT	START COLOUR	TERMINAL No.	OUT COLOUR	TERMINAL No.	OUT COLOUR	TERMINAL No.	OUT COLOUR

TRACK WIRING

go along – have a look at the end for the completed record.

POWER CONNECTION TO THE TRACK

Most proprietary manufacturers have their own method of connecting the cables to the track; a few examples are shown in the illustrations below. The aim of this exercise is to get power on the rails all the way round the track, and the quickest way to check this is to put a locomotive on the track and run it round the circuit. There should be no stuttering or stopping, and when you switch the forward/reverse switch to reverse, the loco should run round the circuit backwards.

Fishplates connect one rail to the next. If you have a problem with no power somewhere, check the fishplates first.

Next check for dirt on the track, as this could stop the power getting to the wheels and subsequently to the loco motor. There are a number of track cleaners on the market specifically for this job.

The first connector is the Hornby R602, as shown in Figs 7.2 and 7.3; it is fitted through slots in the rail between the sleepers and the rail, as shown in Fig. 7.4.

Another type are Peco connectors.

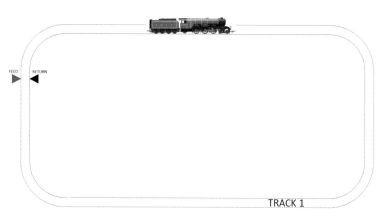

Fig. 7.1 Oval layout with symbols to show connection of the speed controller to the track.

Fig. 7.2 Hornby speed controller connector R602.
HORNBY

Fig. 7.4 The two slots between the sleepers to take the power connector.

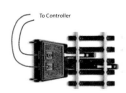

Fig. 7.3 R602 with cables connected, and the unit connected to the track.

Fig. 7.5 Peco ST-273 power connector. PECO

Fig. 7.6 *Wires soldered directly to the rails.*

The second type (see Fig. 7.5) also slides under the track and clips over the base of the track; they are then screwed down to the baseboard, and the wires plug into the rear of each connector. Be careful with the long connector, as this slides under both rails and clips to the furthest rail. Make sure it does not touch the inner rail as this will cause a short.

The third type (see Fig. 7.7) is wire-soldered to the track joiners (fishplates), and can be used anywhere on the layout by replacing the existing joiners (fishplates).

Another way to connect power to your track is by soldering wires directly to each track (see Fig. 7.6). However, you need to look out for a number of things if you are going down this route:

- Clean the part you are going to solder with some emery paper. This could be the side or the bottom of the track
- Remove the plastic sleeper from under the track, as this is going to melt and will make it harder to solder the wires to the track
- Tin the part first: that is, heat up the section and lay down some solder
- Bend the end of the wire to the shape you require before soldering so that it sits on the side of the track and then goes through the baseboard

Fig. 7.7 *Peco PL-81 power connectors.* PECO

- Solder the wire to the track, do not solder to the inner side of the track as that is where the wheel flange goes; repeat for the second wire
- Check for continuity between the track and the other end of the wire. There is such a thing as a 'dry joint', where the joint looks all right but there is no electrical connection because the solder is stuck to the track with flux, not solder

If you are not proficient at soldering, keep to the commercially available connection units.

We have now connected power to the track via a speed controller, which also has forward and reverse switching. Test the track by running a loco,

TRACK WIRING

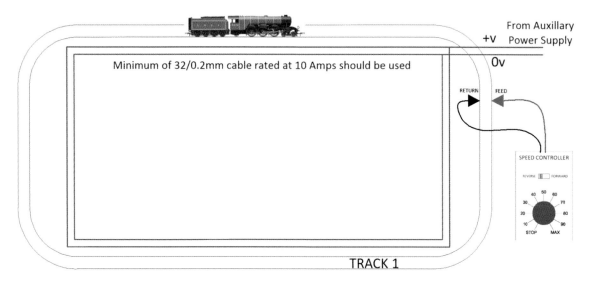

Fig. 7.8 Oval layout with power busbar on the underside of the baseboard. BRIMAL

and check for stoppages and stuttering. If there are none, all is good.

FITTING A POWER BUSBAR

Before we go any further this is a good time to fit a power busbar. It comprises two wires that run around the layout under the baseboard, providing a positive and negative supply to any part of your layout. Busbars are normally run with 32/0.2mm cable as they can end up taking large amounts of current. There are kits available that give you 10m of red and 10m of brown wire, self-adhesive cable ties (*see* Fig. 7.8), and cable spur clamps.

Cable clamps (Fig. 7.9): Use cable clamps to hold the two wires to the baseboard, at approximately 300mm centres. They come with self-adhesive pads to stick to the baseboard. Simply place the two wires between the flaps and then bend them over the wires.

Quick splice connectors (Fig. 7.10): The advantage of quick splice connectors is that when you want to spur off you don't have to cut the busbar: simply add the spur wire to the fitting and clamp the two wires together, as shown in Fig. 7.11. The power for this is normally taken from an auxiliary output on the speed controller.

Power distribution board (Fig. 7.11): The busbar wires come into the board on the left, and leave it on the right. This gives eighteen positive outlets and eighteen negative outlets. The indicator shows green for correct polarity, red for incorrect polarity, and orange for DCC or AC power.

Fig. 7.9 Aluminium self-adhesive cable clamp.
JPR ELECTRONICS

84 TRACK WIRING

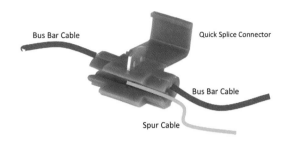

Fig. 7.10 *Quick splice connector.* JPR ELECTRONICS

Fig. 7.11 *Busbar distribution board for positive and negative tack-offs.*

POINT MOTORS

There are three basic types of point motor: the solenoid motor, which in fact is not a motor but two solenoids giving a push/pull effect; the stall motor, which is a motor that simply stalls at the end of its travel; and the servo motor, which is also a motor with a limit on its movement before it switches off.

SOLENOIDS

The most important thing to remember about solenoid-type point motors is that they only require a short pulse of power to energize them. *Any longer than two seconds and the coils will burn out.* A few examples of solenoid-type point motors are shown in the following illustrations.

Each of these point motors requires three wires, as shown in the drawing – Fig. 7.13 – below. Black is shown as the 0v negative, which is common to

Fig. 7.12 *A range of point motors: Peco on the left, Hornby in the centre, and Gaugemaster on the right.*

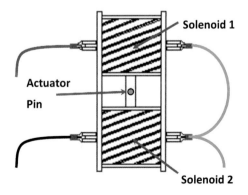

Fig. 7.13 *Cross-section of a twin solenoid point motor.*

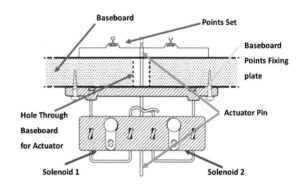

Fig. 7.14 *Cross-section of the fitting of a twin solenoid point motor.*

TRACK WIRING | 85

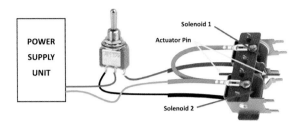

Fig. 7.15 Basic wiring of a twin solenoid point motor.

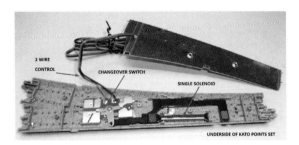

Fig. 7.16 The internal parts of a Kato single solenoid point motor. KATO

both coils. Red is the positive power to one coil, and green is positive power to the other coil.

NOTE: It must be stressed here that not all proprietary manufacturers use this colour coding, so before you connect any power check the manufacturer's instructions.

As you can see from the drawing – Fig. 7.13 – the motor consists of two solenoids – 1 and 2 in the drawing – with a steel bar running through the centre. In the middle of this bar is the actuator pin, which is connected to the points lever. The green wire is the common negative 0v from the power source. The red wire actuates solenoid 1, which will pull the actuator pin towards solenoid 1. The black wire actuates solenoid 2, which will pull the actuator pin towards solenoid 2. Both these actions require a short pulse of power of no more than one or two seconds. The actuator pin in turn moves the points in or out.

SINGLE SOLENOID POINT MOTOR (Fig. 7.16)

This motor is sometimes referred to as a reverse polarity points switch. There is a manufacturer that uses a single solenoid to change a set of points: Kato have built the solenoid into the base of the points, so you don't see the motor. This is a polarity sensitive coil, but being a coil it only requires a short burst of power to energize it. This means that a passing contact switch is required. As mentioned before, passing contact switches have an inherent fault, so we will use a centre-biased toggle switch. This switch needs to be a double pole centre off (DPDT) switch.

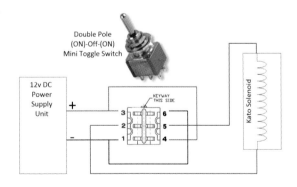

Fig. 7.17 Wiring circuit for the reversing polarity for a single solenoid point motor.

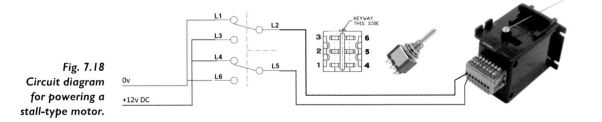

Fig. 7.18 Circuit diagram for powering a stall-type motor.

STALL OR SLOW-MOTION MOTORS (Fig. 7.18)

Tortoise motors are usually more expensive, but they do give a nice slow and realistic point movement. They work by reversing the polarity to the motor, which stays on the motor at all times. They need a DPDT locking toggle switch, and are usually fed from a regulated DC supply. Stall motors can be analogue or digital. Analogue versions operate, depending on make, nominally from a 9 to 12 volt supply DC, and draw little current to move over or when stalled at the end of their travel, typically around 10 to 30 milliamp.

POINTS

There are many different types of points on the market, and it would be impossible to cover all of them in this book. The basic principle is the same for most two-track points: to change the travel direction from one track to another. As seen above, point motors can be surface mounted, or mounted under the baseboard. The anatomy of a set of points is shown in the following image (Fig. 7.19).

The tie bar is what the motor connects to from the underside through the actuator pin hole or the upstands, one at each end of the tie bar. The two closure rails are fixed to the tie bar, so when it moves, they move to change the direction of travel.

The following diagrams are for solenoid-type points, irrespective of their mounting position.

Single point set (Fig. 7.21): Here we show a power source of 12v DC; the switch is a single pole (on)–off–(on) switch. The CDU is optional: if this circuit is reliable then you probably don't need it, however if you get some misfires then it is easy to add one in. (*Although we have colour coded the wires, please check the manufacturer's data.*)

Dual point set (Fig. 7.22): The next drawing shows one switch activating two point motors. Please note that coil 1 is activated on both motors at the same time, so both points will go from ST to

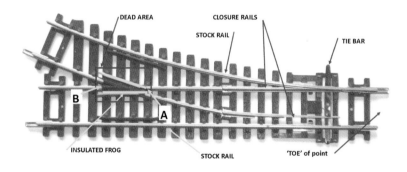

Fig. 7.19 The names of the different parts of a point set.

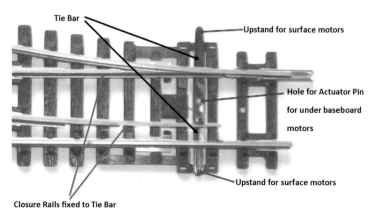

Fig. 7.20 Detail of the actuator part of the point set.

TRACK WIRING

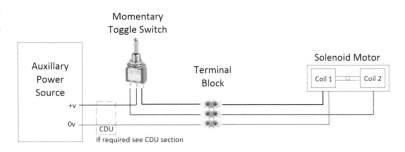

Fig. 7.21 Circuit diagram of a single point set controlled by one toggle switch.

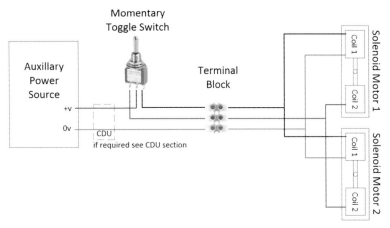

Fig. 7.22 Two point motors controlled by one toggle switch.

Fig. 7.23 Examples of different capacitor discharge units (CDUs).

TU at the same time. I mention this because if you look at Fig. 1.20 in Chapter 1, The Layout, you will need to change P1 and P2 at the same time – if you don't you will have an accident. You also need to check the coils for each point set as you want them both to change to the correct state at the same time. If you get the coils mixed up you will change one set and the other set might not change.

CAPACITOR DISCHARGE UNIT (CDU)

We have mentioned the CDU a few times in this section, so I think it is time to explain what it is and what it does. As mentioned before, most types of point motor are not in fact motors, but are solenoids. When two solenoid coils are placed end to end, a forward and a reverse motion can be created. A solenoid requires a short, sharp punch of current to switch them. These high currents can damage switches, and indeed cause points motors to overheat and burn out. A capacitor discharge unit gets over this problem by supplying this high current as a high energy pulse to the solenoid, which in turn operates the points. The purpose of capacitor discharge units is to store up electricity in a capacitor. On throwing a switch, the electrical charge is released as a pulse of a much larger current.

A capacitor discharge unit is normally powered with AC, generally a voltage of around 16 volts. Note that whether you power the CDU from AC or DC, the output of the CDU will be DC. The

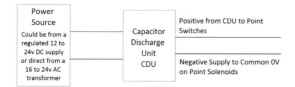

Fig. 7.24 Basic wiring of a capacitor discharge unit from a power source.

CDU may be operated from any AC voltage up to 24 volts. Using a high voltage increases the power from a CDU. Thus operating from 24 volts instead of 12 volts gives four times the power.

WHY YOU NEED ONE

All model rail layouts have a power source, and this power source has a limited capacity. Not all layouts will require a CDU unit. The best way to judge this is, if your points have problems switching over, then you probably need a CDU. The larger the system, the more drain on the power source – also the further away the points are from the power source, the greater the voltage drop. The point motor requires a precise voltage and current to energize it, and the CDU achieves both of these. As this pulse is short it avoids the problem of motor burn-out: the current burst is over by the time the switch contacts open, thus eliminating the back EMF, which can weld the switch contacts together. After the pulse has fired, the CDU will not allow recharging of the capacitors until the switch reopens or is released. Then the cycle can start again with the capacitors recharging.

WIRE SIZE

The CDU delivers a short, sharp jolt of power to the motor, and therefore requires the appropriate wire size. Smaller diameter wire suffers from two problems, namely voltage drop due to its diameter and/or distance from the CDU to the points, and this could affect the performance of the point motor. You should wire from the CDU to the points in either 16/0.2 or 24/0.2 size wire. If you are changing two points at the same time you will be drawing approximately 3 amps at that point in time, and the standard 7/0.2 or 1/0.6 wire will not cope with that, so you need a larger wire size. The short piece of wire the motor may come with should have no effect on the overall performance, but keep it as short as possible when using capacitor discharge units.

BASIC WIRING

The CDU inputs can be connected to any AC power supply up to 24 volts. The most common connection point would probably be the 16v AC auxiliary output usually found on model railway speed controllers. With DC supplies, be sure to connect them with the correct polarity to the CDU. The minimum supply voltage is 12v AC or 15v DC, as below this level the CDU will no longer improve the action of the points. The output terminals are connected to the points switches, with the CDU negative output generally connected to the common centre point of the points motors.

The current from the CDU will make the point motor move up or down depending on which coil is energized. Some point motors have four terminals (wires), because each coil has two wires. Some point motors, such as Seep and Hornby, have three because the manufacturer has connected together one end of each coil to give a common return connection.

MULTIPLE MOTORS

Fig. 7.29 shows the connection of multiple point sets. The wiring is basically the same for one as it is for many, just follow the blue and green wires on the drawing. As you will see, the green wire is the common 0v, and it runs between all the points and then back to the CDU. This is a good case for an 0v (negative) busbar, which will be either 24/0.2 or 32/0.2 wire, and saves running a separate negative wire to each set of points. You cannot run the positive (blue) on a busbar, as each set of points requires two separate wires.

MULTIPLE AND CROSSOVER POINTS

Fig. 7.22 shows a single switch used to activate two sets of points: this would be in a crossover situation, or in the fiddle yard. In this situation you are changing two points at the same time so you will be drawing approximately 3 amps at that switch – the standard 7/0.2 or 1/0.6 wire will not cope with that current, so you need a larger wire size, 16/0.2mm or 24/0.2mm. The short piece of wire the motor may come with should have no effect on the overall performance, but keep them as short as possible when using a capacitor discharge unit: use the three-way terminal block shown in Fig. 4.9 (Chapter 4, Wire and Cables).

Do not use CDUs on motors such as stall motor points, servo motor points or tortoise motors.

Fig. 7.26 Track using isolation fishplates.

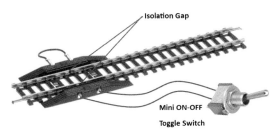

Fig. 7.27 Circuit diagram showing a switch used to switch a section of track on/off.

Fig. 7.25 Point set in straight-through position showing rail polarity.

TRACK ISOLATION

In a DC system track isolation is important, otherwise all trains will move when the speed controller is used. The most common way of isolation is via the points – as can be seen in Fig. 7.25, the yellow rail is away from the blue rail, so cannot supply power to the top right-hand track.

A more permanent way to isolate a section of track is to use plastic fishplates. These are used to join two track sections together with a 1mm gap.

Most proprietary manufacturers produce plastic fishplates. The reasoning behind this could be that you have a section of the layout where you have a completely separate speed controller, and want to use it while the main lines are also being used.

Another way is to use a track isolator section, as shown in Fig. 7.27 (above). Here the left and right rails have been separated, and on the top rail a loop of wire is used to continue the circuit. On the lower rail a switch can be used to energize or isolate the right section of track. Try to ensure that you switch the positive rail rather than the negative rail.

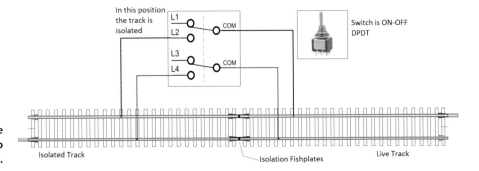

Fig. 7.28 Double pole switch used to isolate both rails.

90 TRACK WIRING

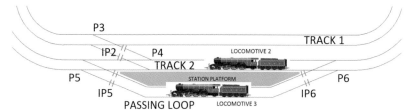

Fig. 7.29 Layout of a passing loop.

In the example on the previous page (Fig. 7.27) only one rail is being switched to create the isolation. In the following drawing (Fig. 7.28) the isolation is controlled by a double pole double throw (DPDT) switch and using plastic fishplates. The connections to the track can be made by soldering directly to the track, or a power connector clip (as shown above) can be used both left and right of the isolation point.

PASSING LOOPS (Fig. 7.29)

As discussed in Chapter 1, The Layout, passing loops can be placed anywhere on your layout, but are most commonly found in stations and shunting yards. A passing loop allows you to run two trains on the one track by placing one in the passing loop. This is achieved by the isolating points P5 and P6. When the points are in the ST position the passing loop is isolated, and when the points are in the TU position the main line section is isolated. Note P3 and P4 will also need to be in the TU position, otherwise you could get power from track 1. You can wire this passing loop in one of two ways: one switch for both points, or one switch for each point set.

One switch for both points: For the wiring diagram see Fig. 7.21. This is the best way of switching a passing loop, as both point sets P5 and P6 are switched at the same time, thereby completely isolating track 2 in the station section. If you only switch P6 to bring the loco into the loop, power is still on track 2 in the station via P5. So any loco in the station will move at the same time as the loco coming into the loop, until you remove power by changing P5. This will then isolate the station section of track 2.

One switch for each point set: For the wiring diagram see Fig. 7.30 below. This circuit shows how to wire the points for individual switches. The use of this system is not recommended until you are more experienced.

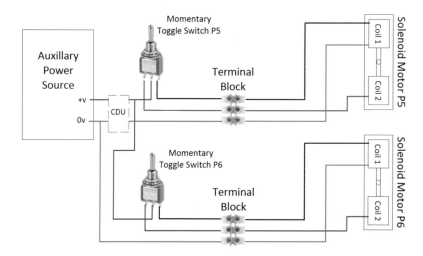

Fig. 7.30 Diagram showing points P5 and P6 with separate controls.

TRACK WIRING | 91

Fig. 7.31 Visual indication of rail polarity.

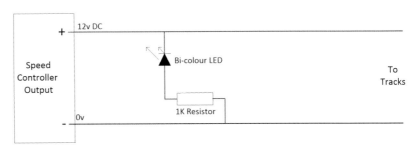

FORWARD/REVERSE SWITCHING

Most speed controllers come with a forward/reverse switch, but not all have an indication. Sometimes it is useful to have a visual indication of the forward or reverse status. This circuit above is very simple and can be added anywhere to give you a visual indication. Green will be forward, red will be reverse.

If your controller does not have a forward/reverse switch, the following circuit (Fig. 7.32) will. It can also be used in sections of track where you want local forward or reverse, such as shunting yards. If you want an indication with the switch you will require a three-pole single throw switch (3PST) for forward/reverse, or a three-pole double throw switch (3PDT) for forward–off–reverse.

REVERSING LOOP (Fig. 7.33)

Reversing loops are used to change the direction of the train and put it back on the original track but going in the opposite direction. As you can imagine, the polarity of the main line will need to change as the train comes down to the points P13 near position IP15. Here we can see that the red rail has gone from the bottom of the image to the top of the image, and at IP15 it is now connecting to a blue rail. However, this cannot happen, and we need to change the main track polarity to red on top and blue at the bottom. This can be done in one of two ways:

- Automatically, using a magnetic sensor on the track and a magnet on the train, which changes the state of the points and the polarity as the train passes over the sensor
- Manually, with two toggle switches

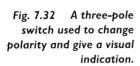

Fig. 7.32 A three-pole switch used to change polarity and give a visual indication.

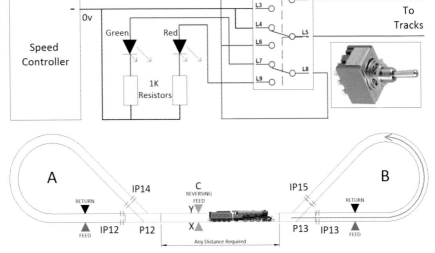

Fig. 7.33 Layout of a reversing loop.

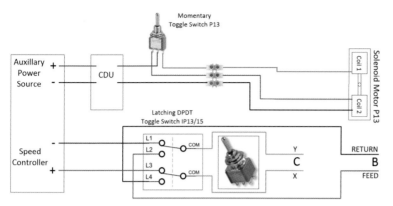

Fig. 7.34 Manual setting of points and reversing polarity.

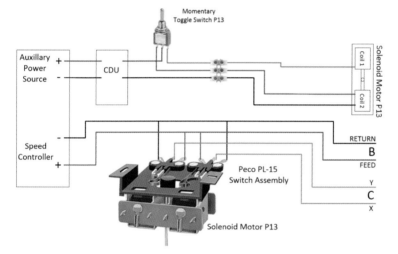

Fig. 7.35 Manual setting of points and automatic setting of polarity change.

In this circuit you need two switches because the one driving the points must be momentary, and the one changing the polarity must be latching. Here we have the power going through a CDU to the momentary centre-biased toggle switch for P13. The second switch is a double pole double throw (DPDT), which changes the polarity at connection point C. You need to keep the polarity at B the same at all times, so that comes directly from the speed controller. The permanent power goes as far as IP13 on the main track, and as far as IP15 on the loop. IP13 and IP15 are track isolation fishplates, as described in Chapter 1, The Layout, Fig. 1.21. The basic problem with this switch arrangement is that you must remember to activate both switches at the same time: this is done manually with one momentary toggle switch and two micro switches (leaf switches) fitted to the points.

In this option you will require a Peco PL-15 switch plate. The momentary switch wiring is the same as Fig. 7.35, however the PL-15 switch plate is fitted on to the point motor (see the manufacturer's instructions). As you change the direction of the points with the momentary switch you also change the polarity of the main track. The speed controller supply is wired directly to connection B feed and return; at the same time these wires are connected to the leaf switches, as shown in Fig. 7.35. Connection C to the main track is wired from the leaf switches.

Whichever circuit you decide to use, you will need to repeat it for loop A. Depending on how far apart the loop points are, the challenge is remembering the position of the points as the train arrives. To solve this problem, see the section 'Points Status Memory' below.

TRACK WIRING

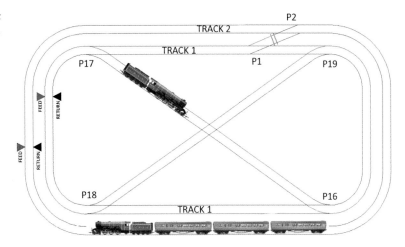

Fig. 7.36 A figure-eight reversing loop layout.

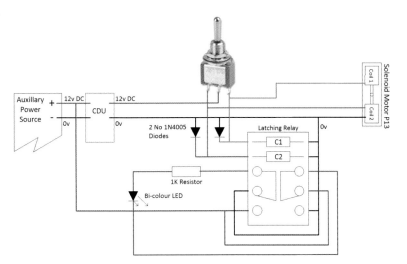

Fig. 7.37 Circuit diagram of points status indication.

FIGURE-EIGHT REVERSING LOOP (Fig. 7.36)

In this layout you have four sets of points controlling two loops: P16, P17, P18 and P19. If you set all points to TU the train will continue in a figure eight until you change it. The *feed* to the track must be as shown: if you change the position of the 'feed and return' there could be problems. The switching of the points is exactly the same as shown in the 'Point Motors' section earlier in this chapter, depending on what type of points you are using.

POINTS STATUS MEMORY (Fig. 7.37)

There is a solution to this problem. The momentary switch can be replaced by a momentary switch with a latching relay. The latching relay switches a bi-colour LED, so you can set the 'straight-through' to green and the 'turnout' to red. The advantage of this circuit is that the latching of the relay remains in that state even when power is removed. The next time you switch on your layout the relay/LED will indicate the position of the points.

A number of manufacturers make this latching unit; a few examples are given in Fig. 7.38.

Fig. 7.38 Examples of points status indication circuits. BRIMAL/GAUGEMASTER

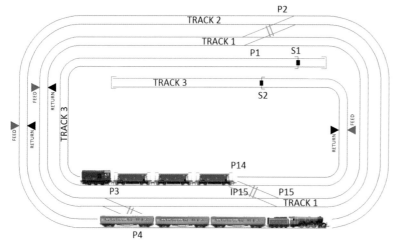

Fig. 7.39 Layout of track with three automatic forward/reverse switches.

AUTOMATIC FORWARD/REVERSE SWITCHING: SHUTTLE

In some cases you may want a train to do its own thing, either in the centre or in some other place on your layout. It would normally be a length of track that goes, for example, from a factory to a depot and back again, and this could be continuous until you want to change it. In this situation you need to change the polarity of the rails so the train reverses back to the start point and changes the polarity, so it can start the journey again. In the layout above (Fig. 7.39) we have added a third track accessed by P14 and P15. The junction of these two points sets will require isolation fishplates, as you are going to run this from a separate speed controller, independent of the others.

You will also notice we have added two sensors, S1 and S2. The position of these will depend on the position of the magnet under the loco, the length of the train, and which end of the loco is closest to the buffer stop. In this example the sensor S1 is close to the buffer stop as the loco will be arriving first, whereas S2 is some way away from the buffer stop as the loco will be reversing into this position and the wagons will be in the siding first. In both cases the magnet under the loco needs to activate the switch before the train hits the buffers.

In the circuit (Fig. 7.40) we have stretched out the track into a straight length for ease of explanation. S1 and S2 are both encapsulated reed switches about 10mm long; they can be placed between the rails as shown, or lengthways. They switch the relay from one state to the next, which in turn changes the polarity of the rails. The speed controller is fed into the relay as shown, and set to the desired speed. The two diodes are there to stop any EMF from the relay coils. The two orange wires feed the rails as shown, and can be connected to the rails in a number of ways, as described at the beginning of this chapter.

There are a number of manufacturers who supply this as a complete circuit – see Fig. 7.41.

TRACK WIRING | 95

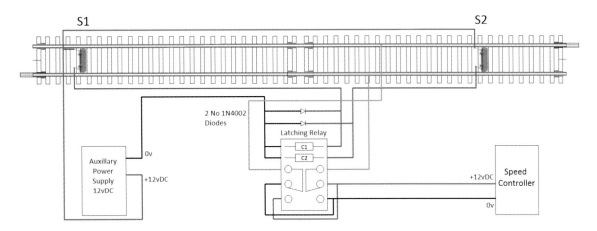

Fig. 7.40 Circuit diagram of an automatic forward/reverse switch using sensors in the track and magnets on the locomotives.

Fig. 7.41 Examples of automatic forward/reverse circuits. BRIMAL/GAUGEMASTER

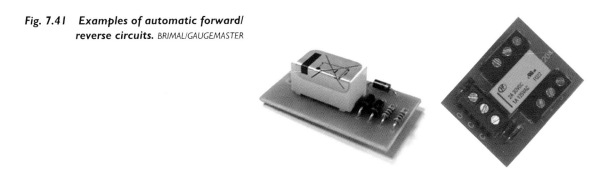

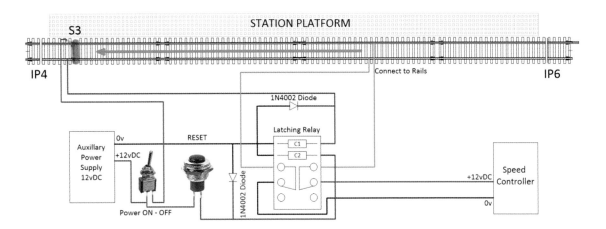

Fig. 7.42 Wiring diagram of a manually operated station stop.

STATION STOP (Fig. 7.43)

The circuit described in this section is a basic semi-automatic station stop circuit. The first thing to do is to isolate the rails in the station section with isolation fishplates at IP4 and IP6. The sensor S3 is positioned near the left side of the platform so that you get the full length of the train in the station, not too close to the isolation point IP4. Depending on the train's speed there will be a little over-run, though you don't want it to over-run on to the powered section of track past IP4. The speed control is fed through the normally closed contacts of the latching relay. When the loco passes over S3 the reed will switch on the first coil on the relay, which will disconnect power to the rails.

The train will remain in the station until the reset button is pressed. This will energize the second coil in the relay putting power back on the rails, and the train will leave the station. If this is on a circular layout, it will happen every time the train comes into the station, unless you switch the power off at the on–off switch. The layout will work fine with this feature switched off – just press the start switch before switching off. This resets the latching relay to the normally closed position – power to the rails.

AUTOMATIC TIMED STATION STOP

Here we have added a timer to the above circuit so that the system becomes fully automatic. The train will arrive in the station and stop in the normal way as above, and this will start the timer. At the end of the time period the latching relay will be reset and the train will move out of the station. The timer can be set from five seconds to three minutes. The timer used is called a monostable timer: this means it has one time pulse, as opposed to astable, which has many pulses. The timer requires a negative pulse through S3 to the start terminal to initiate the time cycle. In Fig. 7.43 the sensor S3 switches the negative line from the power supply – follow the black and blue lines.

If your timer requires a positive pulse to initiate the time cycle, follow Fig. 7.44.

DIAMOND CROSSINGS (Fig. 7.45)

Crossover or diamond crossings are available in left or right, and require no special wiring. It is important to note that on some makes there are three dead areas, so the smaller locos may have problems at slow speeds. Depending on where they are placed on your layout it might be an idea to add some feed and return connections near to the crossover.

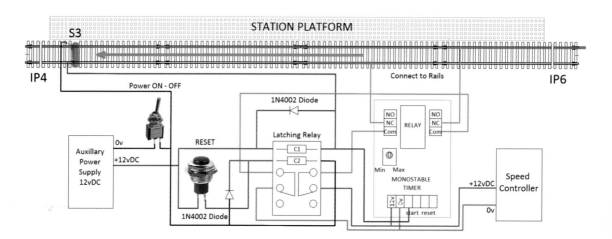

Fig. 7.43 Wiring diagram for an automatic station stop timer with a negative pulse to the timer.

TRACK WIRING

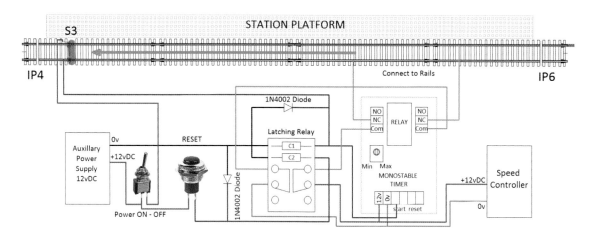

Fig. 7.44 Wiring diagram for an automatic station stop timer with positive pulse to the timer.

Fig. 7.45 Polarity of rails in a diamond crossing.

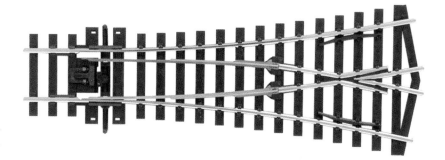

Fig. 7.46 Double slip or 'Y' points. PECO

Y OUTLET POINTS (DOUBLE SLIP) (Fig. 7.46)

Double slip points are wired exactly the same way as a standard set of points, the only difference is that they give you two outlets, each at 12° from the horizontal. For wiring diagrams depending on the type of motor being used, see Fig. 7.15 or Fig. 7.18 above.

THREE OUTLET POINTS (TRIPLE SLIP) (Fig. 7.47)

The triple slip requires two point motors to give you the three outlet option, so you will require two toggle switches, one for each motor – see Fig. 7.48.

In the circuit shown in Fig. 7.48 we are using coil-type motors, so require momentary switches. The sequence of operation is coil 2A energized and coil 1B energized momentarily, which will give you outlet 2. Coil 2A energized and coil 2B energized will give you outlet 1. Coil 1A energized and coil 1B energized will give you outlet 3.

In the circuit shown in Fig. 7.49 we are using stall-type motors, so a double pole switch is used to reverse the polarity of the motor. The sequence of operation is similar to that described above (see Fig. 7.47) – the only difference is that the switch lever will stay in the up or down position. For an indication of which outlet is available, see the circuit described in Chapter 10, Detection.

TURNTABLES (Fig. 7.50)

Turntable wiring is a little complicated. You have three sections of track that need to be controlled,

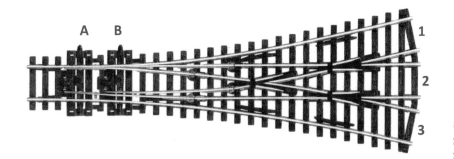

Fig. 7.47 Triple slip points or three outlet points.

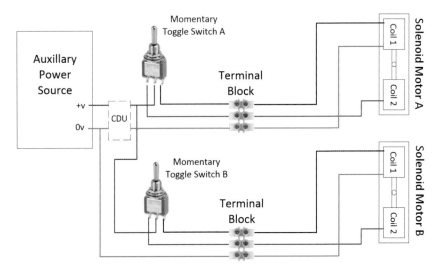

Fig. 7.48 Circuit diagram for triple slip points using coil-type motors with momentary switches.

TRACK WIRING 99

Fig. 7.49 Circuit diagram for triple slip points using stall-type motors with latching switches.

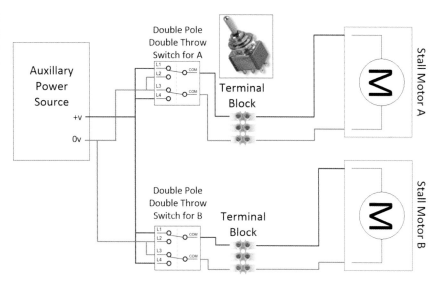

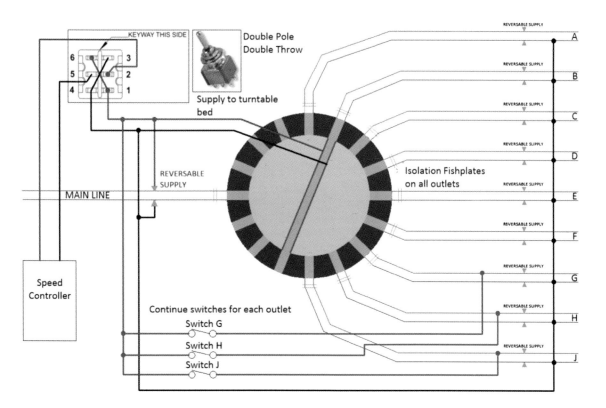

Fig. 7.50 Polarity switch for the turntable bed and outlets.

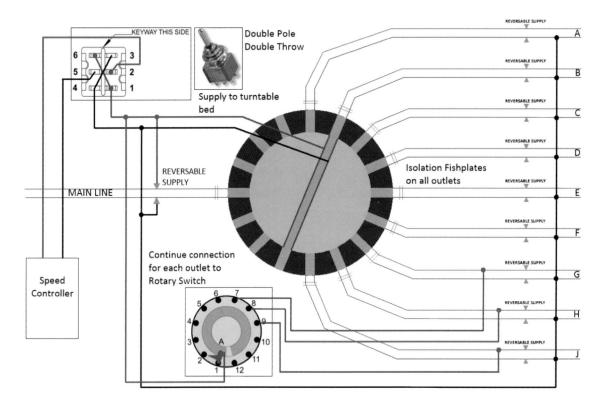

Fig. 7.51 Polarity switch for the turntable bed and rotary switch for outlets.

and the main line coming into the turntable needs to be a reversible supply. Make sure that the left-hand side of the main line does not conflict with any other polarity further down the line. This can be solved by installing insulated fishplates further down the main line, say at the points feeding into the turntable line. The next section is the turntable bed, which needs forward reverse capabilities, and to be isolated from the main line and sidings (outlets).

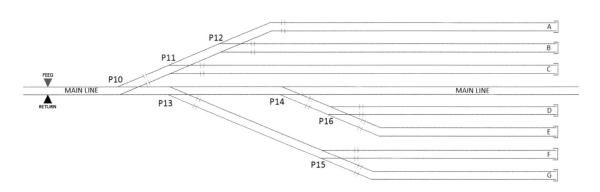

Fig. 7.52 Example of a fiddle yard.

Finally each siding needs to have forward reverse capabilities and an on/off switch. Note the isolation fishplates at the beginning of each siding. (This may or may not be necessary, depending on the type of turntable you are using: see the turntable literature.) You will also have a rotation switch with the turntable, which energizes the motor that drives the turntable bed from one position to another.

In the circuit shown here we are not showing the turntable motor, as this comes with the unit. You will need a single pole switch for each siding you are using. This is just to switch on/off the positive feed to the rails.

FIDDLE YARDS, DIODE MATRIX

A diode matrix is simply a set of diodes placed in an order between a switch and a set of points. You will remember that a diode only allows current to pass in one direction, so we can use this ability to set a number of points to change in order to create a path for the train to travel, all from one switch. As you can see from the drawing (Fig. 7.52), you have sidings A to G all feeding from the main line. In each case there are several points that need to be set for each siding.

For example, to get the locomotive from the main line to siding A, the following must be true: PM7 must be turnout (TU), PM8 must be straight through (ST), and PM9 must be straight through (ST). For this to happen, three sets of points need to be activated from one switch – this is easier than remembering three switches, and which ones. If you now look at switch A row in the table on page 102 you will see that there are three 'Y's: this means that three sets of points need to be set for this to work as described above. If you now look at Fig. 7.53 and follow row 'E' you will see there are three connections starting from this row, one of which is a link (you could use a diode here if you wish).

In the table on page 102 the top line shows the points designation number, the next line has two columns for each set of points ST and TU. On the left-hand side is a column with the siding letter, which is also the control switch letter.

We want to place the loco and wagons in siding A. So the first set of points to be changed to TU is PM7, as shown in the table.

The next set of points to change is PM8 to straight through, as shown in the table.

The final set of points to change is PM9 to straight through, as shown in the table.

PM12 straight through does not have any other 'Y' in the column so you do not need to use a diode: a link wire is placed here. However, using a diode will not affect the circuit in any way, so if you prefer, use a diode.

Repeat this for each siding until you have a 'truth table' similar to the above for your fiddle yard.

Now we need to transfer this information on to a stripboard blank and add the diodes in the correct places. First, mark the top of the stripboard blank with the switch positions and the point motor ST and TU. This will help you get the diodes and links in the right place. Below is the stripboard with the

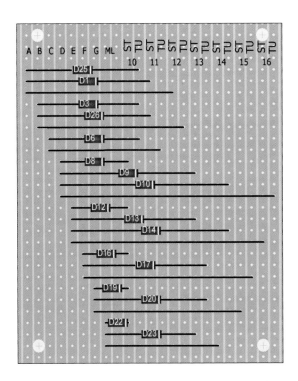

Fig. 7.53 Example of a diode matrix stripboard for fiddle yard in Fig. 7.52.

Points Yes–No Identification

ROUTE	PM10		PM11		PM12		PM13		PM14		PM15		PM16	
SWITCH	ST	TU	ST	TU	ST	TU	ST	TU	ST	TU	ST	TU	ST	TU
A		Y	Y		Y									
B		Y	Y			Y								
C		Y		Y										
D	Y						Y		Y					Y
E	Y						Y		Y			Y		
F	Y							Y				Y		
G	Y							Y			Y			
ML	Y						Y		Y					

diodes and links in place for the example layout above. As you can see, line A on the far left has three connections:

- A diode from A to PM7 TU
- A diode from A to PM8 ST
- A link from A to PM8 ST (if you follow this line down to the bottom you will see no other connections, therefore a link is all right

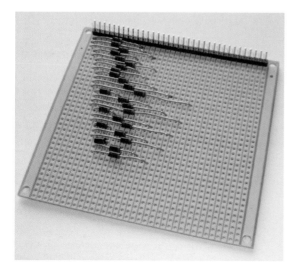

Fig. 7.54 A made-up board based on the truth table above.

Now continue down the stripboard fitting the diodes in the correct places, as in the truth table.

There is one switch called ML (main line). After you have completed your work in the siding you will have to set the main line (ML) back to 'straight through'. The example has three sets of points on the main line, so all three sets must now be straight through, hence the ML switch.

The next stage is to connect the switches and the points to the diode matrix board, as shown in Fig. 7.55 below.

Position the board so the runs of cable are as short as possible.

The switch can be any 'normally open' (NO) contact switch – this could be a push button as shown, a stud connection, or a momentary toggle switch off–(on). One side of the switch is connected to the positive supply voltage, and the other side to the corresponding siding pin.

The points are connected as shown, with one coil to the ST and the other coil to the TU pins. The negative (0v) side of the points is connected to the 0v of your power supply.

CDU: As you are energizing several points simultaneously your power supply is not likely to be able to cope with the power required: it is therefore advisable to connect a CDU, as shown in Fig. 7.55.

TRACK WIRING | 103

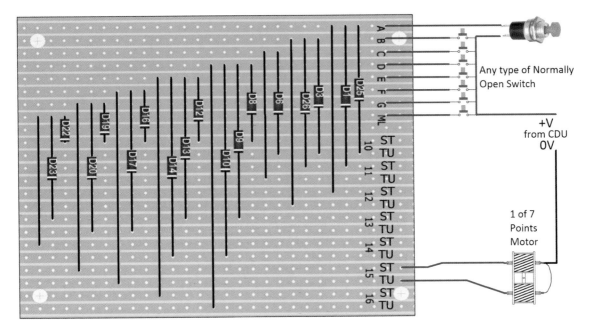

Fig. 7.55 Wiring diagram for a diode matrix board based on the truth table and layout in Fig. 7.52.

UNCOUPLERS

There are four distinct ways to uncouple coaches/wagons: with a manual lever, an electrical solenoid motor, a permanent magnet, or an electromagnet. To add to this there are two types of coupler: the lever coupler and the magnetic coupler.

Fig. 7.57 Example of a magnetic coupler.

Fig. 7.58 Manual lever uncoupler.

Fig. 7.56 Example of a lever coupler.

TRACK WIRING

MANUAL LEVER (FOR THE LEVER COUPLER) (Fig. 7.58)

This is the simplest method, requiring you to move the lever down manually when the coupling is in place. This pushes the bar between the rails up, which forces the couplings apart.

MOTOR DRIVEN (FOR THE LEVER COUPLER)

This method uses a solenoid-type motor to lift up the plate as the coaches go over, forcing the couplings to separate. The wiring for this is shown here. The switch is a momentary toggle switch or a passing contact switch. Both switches will put the

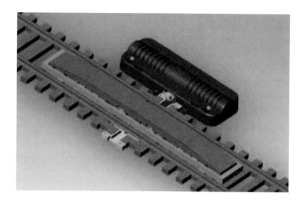

Fig. 7.59 Point motor used to raise the plate to uncouple.

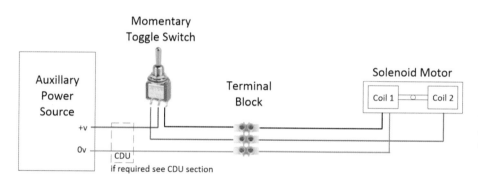

Fig. 7.60 Circuit diagram for the motor-driven uncoupler.

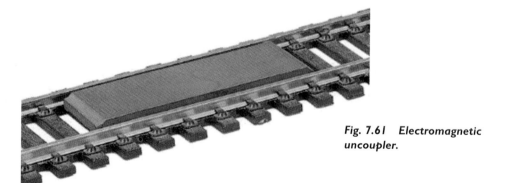

Fig. 7.61 Electromagnetic uncoupler.

Fig. 7.62 Circuit diagram for electromagnetic uncoupler.

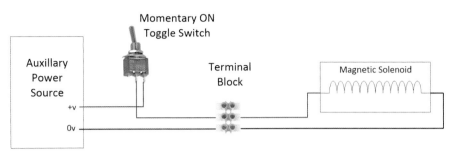

uncoupling bar up, where it will remain until the switch is used again to bring the bar down.

ELECTROMAGNET (FOR THE MAGNETIC COUPLER)

This method consists of a coil under the track, which when energized turns the metal plate magnetic. It requires a momentary on/off switch to energize the coil as the coaches pass over it.

CHAPTER EIGHT

SIGNALS

The railway signalling systems used across the majority of the United Kingdom rail network are line-side signals to advise the driver of the status of the section of track ahead.

SEMAPHORE SIGNALS

During the 1870s, almost all British railway companies standardized the use of semaphore signals, which were then invariably of the lower quadrant type. From the 1920s onwards, upper quadrant semaphores almost totally supplanted lower quadrant signals in Great Britain, except on the former GWR lines. The advantage of the upper quadrant signal is that should the signal wire break, or the signal arm be weighed down by snow (for instance), gravity will tend to cause the signal to drop to the safe 'danger' position. In a lower quadrant signal, the opposite may happen, sending the signal to 'off' when in fact it should be illustrating 'danger'. Their spectacle cases (lens casing), which are on the opposite side of the spindle on which the signal arm is pivoted, must therefore be sufficiently heavy to prevent this happening.

There are two main types of semaphore: stop and distant. The stop signal consists of a red, square-ended arm, with a vertical white stripe typically 9–12in (230–300mm) from the end, and advises the driver as to whether the line immediately ahead is clear or not. A stop signal must not be passed in the 'on' (danger) position, except where specially authorized by the signalman's instruction. By night it shows a red light when 'on' and a green light when 'off' (clear).

The other type is the distant signal, which has a yellow arm with a 'V' ('fishtail') notch cut out of the end, and a black chevron typically 9–12in (230–300mm) from the end. Its purpose is to advise the driver of the state of the following stop signal(s); it may be passed in the 'on' position, but the driver must slow his train to be able to stop at the next stop signal. When 'off', a distant signal tells the driver that all the following stop signals of the signal box are also 'off', and when 'on', it tells the driver that one or more of these signals is likely to be at danger. By night, it shows a yellow light when 'on', and a green light when 'off'.

Current British practice mandates that semaphore signals, both upper and lower quadrant types, are inclined at an angle of 45 degrees from horizontal to display an 'off' indication.

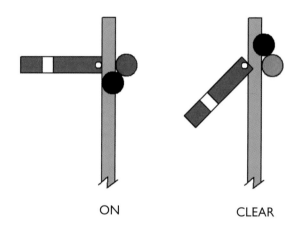

Fig. 8.1 *Semaphore stop signals.*

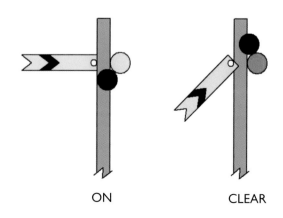

Fig. 8.2 *Semaphore distant signals.*

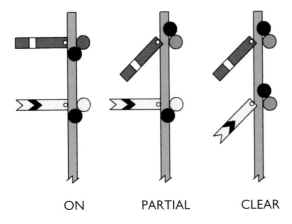

ON   PARTIAL   CLEAR

Fig. 8.3 *Semaphore combination stop and distant signals.*

Most proprietary manufacturers have a semaphore signal of one sort or another. Some are manual signals, although these can be converted to electrical operation; others are designed for electrical operation only.

BASIC SEMAPHORE WIRING (Fig. 8.6)

There is a range of semaphore signals that come with a motor on an extended base, which goes through the baseboard; the hole required is 14mm. The circuit in Fig. 8.6 shows how to connect this signal.

The switch can be either a push button, leaf or reed, or a contact on a relay.

These signals will work with a power source from 9v DC to 14v AC, though it is advisable to check the manufacturer's instructions. The power is connected to the red and black wires, red being +V. The yellow wires must not have power – they energize the motor when they are momentarily joined together. This can be a reed switch on the track, a push button or momentary toggle switch, or a combination of the above. The second time the switch is energized the signal will revert back to its original state.

COLOUR LIGHT SIGNALS (Fig. 8.7)

Colour light signals exist in the UK in two-, three-, four- and five-aspect forms. The main aspects are as follows:

- Flashing green: *Clear* – the train may proceed normally. This aspect was only used on the 140mph (230km/h) trial section of line between Peterborough and York
- Green: *Clear* – the train may proceed normally. (On the 140mph section of line above it means that the next signal will show two yellows)

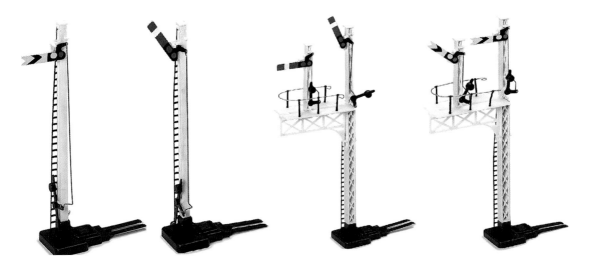

Fig. 8.4 *A range of manual semaphore signals.* HORNBY

108 SIGNALS

Fig. 8.5 *Two types of electrically operated semaphore signals.*

- Double yellow: *Preliminary caution* – the next signal will be displaying a single yellow aspect
- Single yellow: *Caution* – the driver must prepare to stop the train at the next signal
- Flashing yellow: *Caution or preliminary caution – single or double aspect* – warning of a diverging route set at high speed turnouts; the route is set and will give green aspects after turnout
- Red: *Danger/stop*

The single yellow, double yellow and green aspects are known as 'proceed aspects' as they allow the train to pass the signal; the red aspect always requires the train to stop.

TWO-ASPECT SIGNALS (Fig. 8.7)

A simple two-aspect colour light signal would act as a replacement for a semaphore 'stop' signal, or

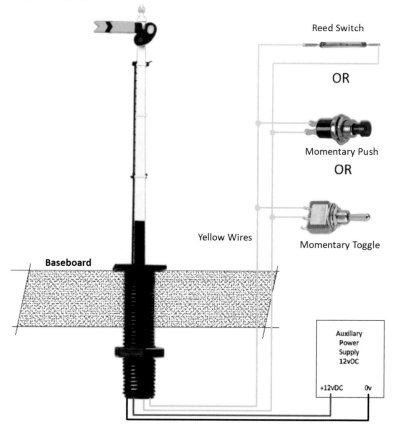

Fig. 8.6 *Circuit diagram to wire an electronic semaphore signal.*

SIGNALS 109

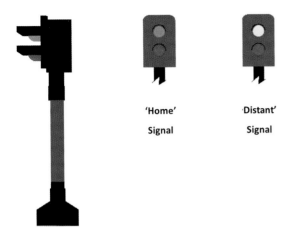

Fig. 8.7 The sequence of a two-aspect electric signal.

a 'home' signal. The 'home' signal would have green and red aspects; the distant signal would have yellow and red aspects.

Two-aspect signals will use either filament lamps or LEDs, and it is important that you know which one you are using as the filament lamps will normally be 12v (they can also be AC or DC), whereas the LEDs will be a lower voltage, around 2 to 3 volts. Next you must check the number of wires coming from the signal unit – it could be two, three or four wires. Most commonly there are three wires: a common negative (0v), one for the red bulb/LED, and one for the green bulb/LED. Apart from the dropping resistor, the wiring is the same, so note where I have put the resistor.

INDIVIDUAL SIGNALS (Fig. 8.8)

A simple on/off miniature toggle switch will be sufficient for each signal.

In the circuit illustrated in Fig. 8.8 the bulbs are LEDs, so a 1K dropping resistor is required as shown; the toggle switch is a simple SPDT on/on switch. If you are using wheat-germ bulbs or Lilliput bulbs then the dropping resistor is not required, as long as the bulb works off the same voltage as the supply voltage. The same circuit can be used for the 'distant' signal by replacing the green for yellow. Note the following:

- The switch can be either a toggle switch, slide switch or a leaf switch, or the switch on some point motors
- The colour of the wires on proprietary manufactured signals may not be red, green or black, so it is important to check the manufacturer's instructions

DUAL SIGNALS FROM ONE SWITCH (Fig. 8.9)

In the circuit shown in Fig. 8.9 both the 'home' and 'distant' signals are controlled by one toggle switch. Here we are using a DPDT (double pole double throw) toggle switch. The left side (terminals 1, 2 and 3) control the 'home' signal, and the right side (terminals 4, 5 and 6) control the 'distant' signal.

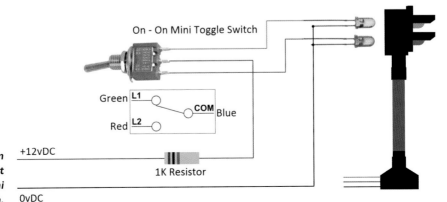

Fig. 8.8 Wiring diagram to operate a two-aspect signal with one mini toggle switch.

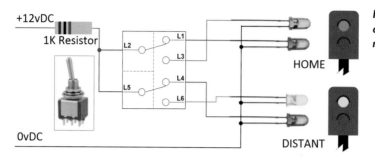

Fig. 8.9 Wiring diagram to operate a 'home' and 'distant' signal from one mini toggle switch.

There is some conjecture regarding the sequence of the signals, so changing the sequence can be done by swapping around pins 4 and 5, or replacing the colour of the bulbs. The principle of the resistor and bulbs applies as above.

Note: LEDs are polarity sensitive, therefore the short lead (cathode) should always be negative (0v), and only work on a small DC voltage – 2 to 3v: hence the reason for the 'dropping resistor'. With wheat-germ bulbs the polarity is not important as they are miniature filament lamps, and rated at 12v DC or AC.

DUAL SIGNALS AT POINTS (Fig. 8.10)

In the example shown in Fig. 8.10, two home signals are being controlled by a set of points. The switch could be a latching relay as shown in Fig. 8.19, or the Gaugemaster switch PL-15 that fits on to the Hornby or Gaugemaster points. When the points are changed at the control station the signals will automatically change to give a 'go ahead' to that track.

TWO-WIRE SIGNALS (Fig. 8.11)

Some LED signals have only two wires to power both the red and green LEDs. An LED is polarity sensitive, so if the polarity is reversed the LED will not work – so if two LEDs are joined together but with opposing polarities, only one will be on at any time. The circuit inside the signal is similar to that shown in Fig. 8.11.

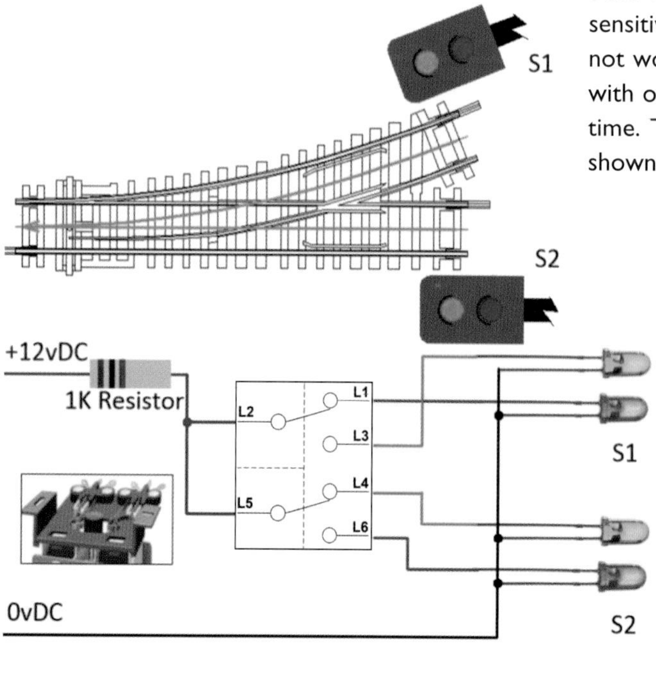

Fig. 8.10 Wiring diagram to operate two signals at a set of points with leaf switches on the point motor. BRIMAL

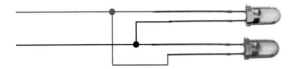

Fig. 8.11 Internal wiring of a two-wire, two-aspect signal.

The principle of this circuit is exactly the same as the circuit we have used to create a forward/reverse switch for the power to the track. We need to use a double pole double throw switch (DPDT) so that the switch changes the output of both the positive and negative. It is advisable to check the manufacturer's instructions, as sometimes the dropping resistor may be incorporated in the signal assembly; if this is the case then you should omit the resistor shown in the circuit in Fig. 8.12. The switch could also be the Gaugemaster PL-15.

THREE-ASPECT SIGNALS (Fig. 8.13)

Controlling a three-aspect signal in model railways can be done with a special toggle switch that has three 'on' positions. It is designated at on–on–on (DPDT). You could also use a rotary switch, however this does take up a little more space than a mini toggle switch. The sequence of operation is shown in the diagram (Fig. 8.13). As the train approaches, the first signal will be green, the second signal will be yellow, and the third signal will be red. Each signal will be at the beginning of a block that is longer than the full length of the train.

The circuit shown in Fig. 8.14 controls one signal manually. The link from 3 to 5 must be put in place first; again, if you are using filament lamps then the dropping resistor will not be required.

In the following circuit we are using a four-pole three-way rotary switch (Fig. 8.15); the advantage of this switch is that you can control three separate signals from the one switch. Circuits A and C can be wired to two other signals, by placing the wires one number further over; the sequence can be maintained for all three signals. The green shown on segment B is in position 2, so in segment C it would be placed in position 3, and in segment A in position 4; it would be the same for the other two LEDs.

GROUND SIGNALS (Fig. 8.16)

GPLS ('ground position light signals') are always illuminated, and are located either on the ground or on a post. Their display is as follows:

- Either two red lights or one white light and one red light in a horizontal arrangement, meaning 'stop'
- Two white lights at a 45-degree angle, meaning 'proceed'. The driver may pass this signal with caution and a speed that allows the train to stop short of any obstruction

Ground signals are available from most suppliers, and because each manufacturer will have their own way of achieving the red/white LED, it is best to see their instructions. If you want to make your own, the circuits shown here may help.

Red/white LEDs are available, and the internal circuit is shown in Fig. 8.17 below. We need to reverse the polarity to get the different colours, so

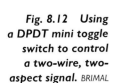

Fig. 8.12 Using a DPDT mini toggle switch to control a two-wire, two-aspect signal. BRIMAL

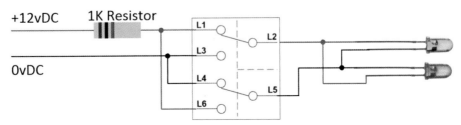

112　SIGNALS

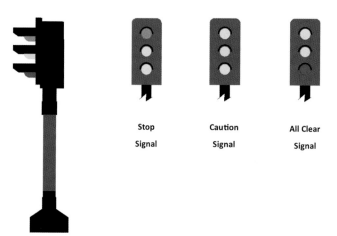

Fig. 8.13　The sequence of a three-aspect electric signal.

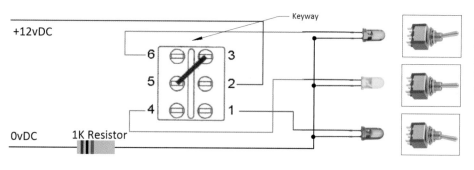

Fig. 8.14　Circuit diagram to control a three-aspect signal with one mini toggle switch.

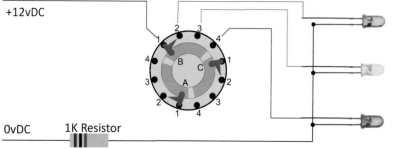

Fig. 8.15　Circuit diagram to control a three-aspect signal with a three-pole, four-way rotary switch.

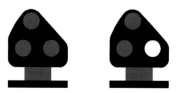

Fig. 8.16　Light sequence of a three-light ground signal.

for red we need one at positive and two at negative, and to get white we need two at positive and one at negative.

In this circuit we are using a three-pole four-way rotary switch to achieve the correct sequence of operation. In position 1 on the switch you will have one and two as red, and three as off. In position 2 you will have one as red, two as white and three as

SIGNALS

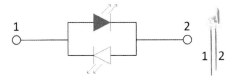

Fig. 8.17 Internal wiring of a dual-colour, two-wire LED.

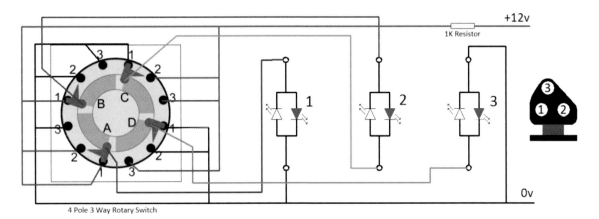

Fig. 8.17a Wiring diagram to control a ground signal to give the correct light sequence using a rotary switch.

off. In position 3 you will have one off, two at white and three at white.

AUTOMATIC SIGNALS

In this layout (Fig. 8.18) signals S1, S2 and S3 are controlled by the point motor switches for P10 and P14. If the main line is being used, then you require S2 to be green and S1 and S3 to be red. If you have a train coming from the top siding, then the points P10 need to be in 'turnout' position, S1 is green and S2 and S3 will be red. Alternatively, if the train is coming from the lower siding, then P14 needs to be in 'turnout' position and the signal S3 is green, and S2 and S1 will be red. The circuit in Fig. 8.19 shows how the points and signals are controlled by the point switches P10 and P14. This system of interlocking points can be expanded to cover all the points and signals on your layout.

There are a number of ways to automate the signals, however all require a sensor of some sort to detect the presence of the train. The sensor can be a reed switch, a micro switch, or IR detectors, but with all of these the sensor has to remain active for a while after the train has passed otherwise the signal just flashes from one aspect to the other and back

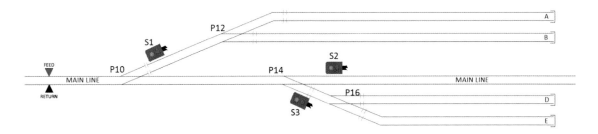

Fig. 8.18 Track and signal layout for sidings off the main line, signals S1 to S3.

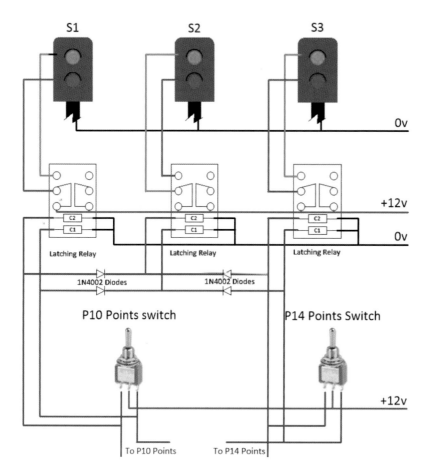

Fig. 8.19 Wiring diagram to switch three signals using two point switches and three latching relays.

again. To make it realistic the sensor has to latch for a predetermined period, either until the train has reached the next signal, or until it has completely passed the signal.

In reality one signal controls the next signal in conjunction with the position of the train, and what the signal controller wants the train to do. Cascading signals is possible, but this does require a certain amount of relay circuits, wiring and sensors.

CHAPTER NINE

LIGHTING PROJECTS

Bringing your layout to life is important, and one of the best ways to do this is to introduce lighting in all forms.

LIGHT SOURCES

The majority of light sources on your layout are going to be from one or more of the following bulbs:

A: 5mm-diameter LEDs. These are available in most colours and brightnesses. Most LEDs require a voltage of 2 to 3 volts DC. However, some are available with a built-in resistor so can work directly from 12v DC.

B: 3mm-diameter LEDs. These are available in most colours and brightnesses. Most LEDs require a voltage of 2 to 3 volts DC.

C: 2mm tower LEDs. These are available in most colours and brightnesses. Most LEDs require a voltage of 2 to 3 volts DC.

D: Bi-colour LEDs are available in 5mm and 3mm. Most require a voltage of 2 to 3 volts DC.

E: Lilliput (wheat-germ) bulbs are 3mm in diameter and are available in most colours; they are usually 12v.

F: Lilliput screw-in filament lamps, LES (E5), are available in clear, red, green and amber, and are either 6v or 12v.

G: Standard torch-type bulbs, MES (E10); they are usually clear, but voltages range from 2v to 24v.

The problem with filament bulbs (E, F and G) is that they generate heat, so care must be taken not to place them too close to plastic or cardboard as it will start to melt or go brown.

Referring to Fig. 9.2, you will need the following holders to take the bulbs described above:

Type 1 for the bulbs categorized in A and D.
Type 2 for those in B and E.
Type 3 for Lilliput E5 bulbs, categorized in F.
Type 4 for MES E10 bulbs, categorized in G.

THE LIGHT OUTPUT OF LEDs

The light output of LEDs is rated in either candelas (also candle) or lux, with the candela being the more common. Both are a measure of luminous intensity. The higher the candelas/lux, the brighter the light. Standard LEDs emit only a modest amount of light, not even 1 candela. They are rated in millicandelas, or thousandths of a candela. Millicandela is typically abbreviated as 'mcd'. The typical intensity of a standard 5mm red LED is about 50 to 75mcd.

Fig. 9.1 Different types of LED and filament lamp. JPR ELECTRONICS

Fig. 9.2 Different types of LED and filament lamp holder. JPR ELECTRONICS

LIGHTING PROJECTS

BASIC LEVEL CROSSING (Fig. 9.3)

A basic level crossing will have a controller using a pulse timer (astable). Power is fed from the 12v DC supply through an on–off switch to the timer. Each left-hand LED is connected to the timer via the blue wire, and each right-hand LED is connected to the timer via the blue wire: this will give a left/right pulse similar to the real thing. The pulse and pause variables should be set for equal 'on' and 'off', so that when the timer is in pause the red wire is connected, switching on the left lamps, and when the timer is in pulse the blue wire is connected, switching on the right lamps.

The 1K resistor is for all LEDs that are low voltage; if you are using LEDs or filament bulbs at 12v then do not use the resistor.

The lights will flash only when the power switch is in the 'on' position.

FULLY AUTOMATIC LEVEL CROSSING (Fig. 9.4)

For this crossing we have added two sensors and a latching relay. The sensor S1 will be triggered by the train and will switch the latching relay, which in turn will start the pulse timer. The wiring for the pulse timer is exactly the same as in the circuit in Fig. 9.3, except the positive power (the purple wire) is coming from the latching relay. When the train

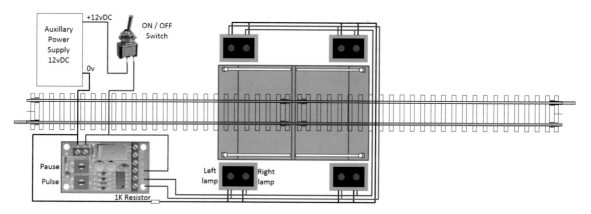

Fig. 9.3 Basic level crossing using a pulse timer. BRIMAL

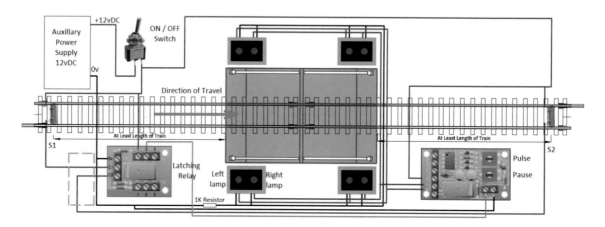

Fig. 9.4 Automatic level crossing working with trains moving from left to right.

LIGHTING PROJECTS   117

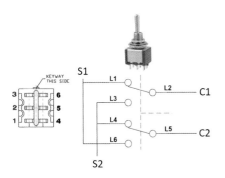

Fig. 9.5 Switch between sensors S1 and S2 to change the direction so the train will switch S2 first.
BRIMAL

reaches the second sensor S2 it will switch the latching relay off, which will switch the pulse timer off. The system is now ready for the next train to energize it.

In the bottom left corner of Fig. 9.4 is a green dotted box. If you add the switch circuit here you can change the direction of travel, so the sensor S2 starts the operation and S1 switches it off. This switch will change the direction that S1 and S2 sense the train, so S2 will become S1, and S1 will become S2: this allows the level crossing to work when the train is travelling from right to left.

LIGHTHOUSE BEACON (Fig. 9.6)

In this circuit you can use either a 5mm LED or a 3mm LED, depending on space restrictions. Both LEDs are available in high brightness: with the 5mm you can get up to 40,000mcd, and with the 3mm you can get up to 20,000mcd. Set the pulse (on) to about one second, and the pause (off) to about five seconds.

WELDING SIMULATOR (Fig. 9.7)

A very simple welding simulator consists of an ultra bright white LED and a standard blue LED – they can be either 5mm or 3mm. The circuit in Fig. 9.7 shows the white LED connected to the NC contact, which is the pause variable, and the blue connected to the NO contact, which is the pulse variable. By adjusting these you can create a very passable simulator, which can be placed inside a factory building. The drawback is that the simulation is repetitive and has no light variation.

ARC FLASH (Fig. 9.8)

The circuit shown in Fig. 9.8 is designed to simulate the arc flash made by electric trains as they cross points and crossings. The flash can be triggered by the train using a magnet on the underside of the train and a sensor between the tracks. The ultra bright blue or white LEDs (or a combination of both) are set into the baseboard so they just protrude to the height of the sleepers. The sensor is set in place as near to the LEDs as possible between non-moving parts of the points. The

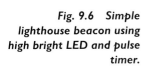

Fig. 9.6 Simple lighthouse beacon using high bright LED and pulse timer.

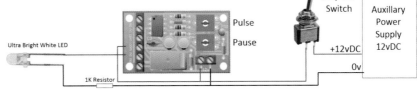

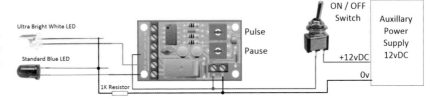

Fig. 9.7 Simple circuit to simulate a welding light using a white and blue LED and a pulse timer.

118 LIGHTING PROJECTS

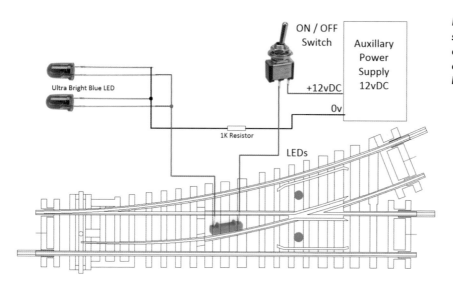

Fig. 9.8 Circuit to simulate the arc flash at points using a sensor and blue LEDs fitted in between the rails.

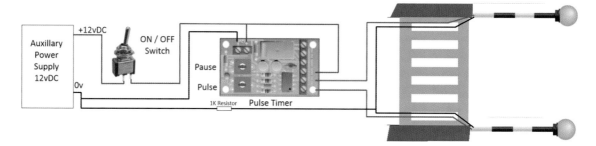

Fig. 9.9 Circuit to drive zebra/pelican crossing lamps using a pulse timer.

ideal situation is that the LEDs flash from under the locomotive.

ZEBRA CROSSING (Fig. 9.9)

In the circuit shown in Fig. 9.9 we are using the timer to run two belesher beacons at a zebra crossing. The beacons are normally supplied with an LED or small filament lamp. Check the voltage they require before you switch them on. If they are 12v then remove the 1K resistor. The circuit below each beacon is wired to the relay separately, so when one is on, the other is off, and vice versa, so set the pause and pulse times to the same length. If you want them both to flash on and off at the same time, then wire them both into either NC or NO of the relay output; this will allow you to adjust the 'on' and 'off' times independently.

STREET LIGHTS

In the circuit illustrated in Fig. 9.10 we show three examples of street lights that are available; however, there are many more. The circuit shows the lights being powered from a 12v DC supply. It is important to check the voltage they require before switching on. If they are LEDs then you may have to put a 1k resistor in the red wire after the on/off switch. Keep an eye on the wattage of these LEDs, as the resistor is probably rated at 0.25 watts. Once you run LEDs totalling more than 0.25 you should increase the size of the resistor to 0.5 or 1 watt. If they are 12v DC lamps then the power supply will become the thing to watch, as everything being run from it must not exceed its maximum rating.

LIGHTING PROJECTS | 119

Fig. 9.10 Simple on/off circuit for powering working street lights.

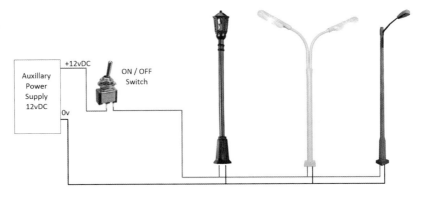

Fig. 9.11 Circuit diagram to control building lighting. Each switch can control up to ten LEDs.

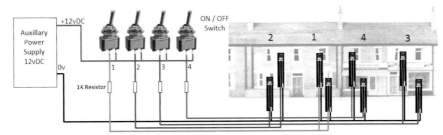

BUILDING LIGHTING (Fig. 9.11)

In the circuit shown in Fig. 9.11 we are using standard brightness LEDs inside the building. The circuit shows an on/off switch for each building, though you could have a switch per street, for example. I would use clear white LEDs in the businesses as they would normally have fluorescents, and warm white LEDs in the houses. Here we have put a dropping resistor after each switch, rather than on the common negative wire: this is just to ensure that we don't overload the resistor with too many LEDs.

CARRIAGE LIGHTING

There are two main ways of fitting carriage lighting. First, it is possible to pick up power from the tracks through the wheels of the carriage. However, on a DC system there is no power on the track when the train is stationary, so the lights will go out unless you have a back-up. Alternatively you can fit a small battery inside each carriage connected to the lights via a miniature switch. This means a lot of batteries and a lot of switches to remember to switch on and off.

PICK UP FROM THE TRACK

To start with, the wheels on most carriages are plastic so you will need to change the wheels to metal, otherwise this modification is not going to work. First decide where you are going to solder the connecting wires to the pick-up fingers: near the fixing screws is the best place, with one red and one black wire. Now install the two pick-up fingers to one of the bogies, as shown in Fig. 9.13. As you can see, the wires have been soldered near the fixing screw in the centre of the pick-up fingers. The wires have been threaded through the fixing spigot so they come out inside the carriage. In Fig. 9.14 the fingers

Fig. 9.12 Wheel pick-up plate kit for carriage lighting.

LIGHTING PROJECTS

Fig. 9.13 Underside of bogie showing pick-up plates fitted.

Fig. 9.14 Topside of bogie showing pick-up plates touching the four wheels.

can be seen to be too long so they need to be cut back so they are no higher than the wheels.

Refit the bogie and ensure the wires come through into the carriage body. Check that the bogie can turn without fowling the wires. Depending on what system you have purchased, these wires will either connect directly to the LED strip, or go through a small electronic circuit. In this coach the bench seat in the last compartment had to be removed to accommodate the circuit. The reason for this circuit is to give continuous light for a short period while the train negotiates 'dead sections' of track, or when it comes into the station to stop. You may recognize the bridge rectifier (the round black object in the top left corner of the PCB): this gives the positive and negative to the lights irrespective of which way around the power is coming to the circuit from the track.

The next operation is to fix the LED strip to the inside roof of the carriage. First, solder the positive and negative wires to the LED strip, and then stick it to the roof. In Fig. 9.18 below I have used glue, but you could use double-sided tape. Ensure the wires come down the same end as the circuit. Now connect the input fingers to the circuit and the LED to the circuit, according to the manufacturer's instructions. Check that the system works by applying 12v DC to the wheels: if all is fine, close up the carriage and it is ready to roll.

BATTERY OPERATION (Fig. 9.19)

There are kits available that run off batteries. The battery holder needs to be mounted inside the car-

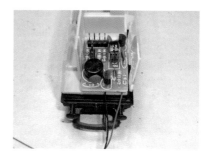

Figs 9.15/9.16 Control circuit shown in plate with pick-up wires ready to be connected.

LIGHTING PROJECTS 121

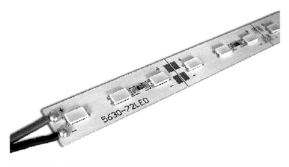

Fig. 9.17 Example of LED tape, and the positive and negative wires soldered to the tape.

Fig. 9.18 The LED tape stuck to the inside roof of the carriage.

riage, and the on/off switch needs to be mounted where it can be easily accessed. Here I mounted the switch on the side of the floor with a cut-out in the side of the carriage. Connection to the LEDs is the

Fig. 9.19 Battery-operated carriage lights showing the position of the on/off switch and battery holder.

same method as above, just check that you connect the red to positive and black to negative.

Note: In this kit the LED strip will require a 3v supply and therefore will not have built-in resistors. Check that it says 3v and not 12v on the strip.

LEDS

There are a few interesting facts about LEDs that may help with your lighting projects. First, all the circuits I have shown have a 1K resistor: if you change the value of this resistor slightly, the brightness of the LED will change. For example, if you fit a 220ohm resistor the LED will be brighter than with the 1K, and if you fit a 2K resistor the LED will be dimmer.

BI-COLOUR LEDS

As the name implies, the LED has two colours – although in fact it has three colours. Bi-colour LEDs come in two-pin or three-pin configurations.

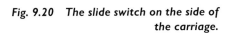

Fig. 9.20 The slide switch on the side of the carriage.

LIGHTING PROJECTS

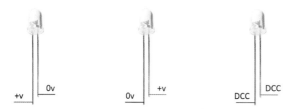

Fig. 9.21 Bi-colour LED connections to give three colours.

Two-Pin LEDs

In Fig. 9.21, the first image shows positive on the long lead: this will give a *red* LED. The second image shows positive on the short lead: this will give a *green* LED; the third image shows AC to both leads: this will give an *orange* LED.

> Note that the dropping resistor has not been shown, but must be there. The voltage to the LED must not exceed its working voltage, normally between 2 to 3 volts.

This arrangement could be used in the forward/reverse switch. Shown in Chapter 7, Track Wiring, this is done by replacing the red and green LEDs shown in grey with a single bi-colour LED. If the colour is not correct for how you want to indicate, just turn the LED round by 180 degrees and reconnect it.

Three-Pin LEDs (Fig. 9.23)

The three-pin LED has one advantage over the two-pin, and that is that you can have two separate circuits driving it. It still has the orange colour when both LEDs are on. The LED has three pins, each one a different length. The centre pin (the longest) is called the common cathode where you connect the 0v supply. This is also a good place to put the dropping resistor, as it saves having to have one on each positive pin. The next longest is the red anode, and the shortest is the green anode.

This LED is ideal for indicating the positions of point motors. It can be wired to a changeover switch on the points. (See drawings in Chapter 10, Detection: Figs 10.1, 10.2, 10.3, 10.4, 10.5 and 10.7.) This arrangement could be used in the forward/reverse switch. It is shown in Chapter 7, Track Wiring, by replacing the red and green LEDs shown in grey with a single bi-colour LED. If the colour is not correct for how you want to indicate, just turn the LED round by 180 degrees, and reconnect it.

I am sure in time you will find many more uses for this type of LED.

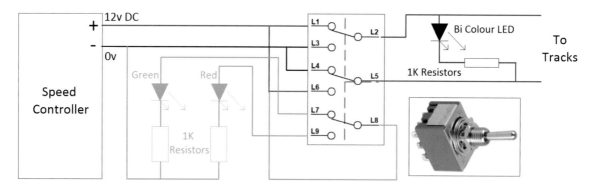

Fig. 9.22 A bi-colour LED used to indicate forward or reverse polarity to the track.

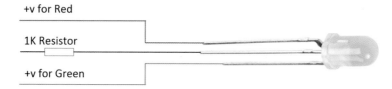

Fig. 9.23 Circuit diagram for a three-pin bicolour LED.

CHAPTER TEN

DETECTION

Detection covers all aspects of your layout. This could be wanting to know where your train is, to wanting to know what position your points are in at the far end of your layout. There is a range of detection devices available: these could be micro switches, spring lever switches, magnetic switches, or infrared switches. In this chapter we will show how to wire most of these in a specific position.

POINT STATUS INDICATIONS

Most manufacturers produce a switch for their points to indicate what state they are in. There are many ways to indicate the status of a set of points; here we will try to show as many as we can, with the relevant circuit.

Fig. 10.1 shows the Peco PL-13 switch, which fits on a range of underboard point motors. The actuator pin fits through the hole in this switch to move the switch from one contact to another when the points are changed. Here we are using it to indicate the position of the points.

Fig. 10.2 shows the Peco PL-15 switch, which again fits on a range of underboard point motors. The actuator pin fits through the hole in this switch to move the two leaf switches from one contact to another

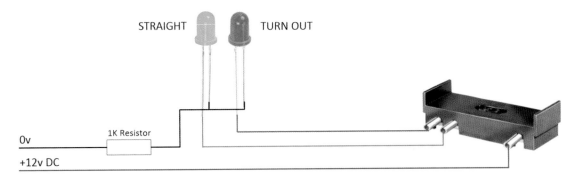

Fig. 10.1 A Peco PL-13 switch mounted on a point motor to give point status indication.

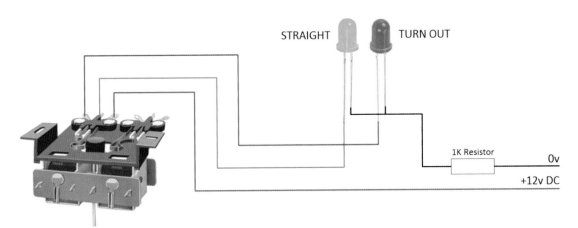

Fig. 10.2 A Peco PL-15 switch mounted on a point motor to give point status indication.

when the points are changed. Here we are using it to indicate the position of the points. The other leaf switch could be used for many other indications.

In Fig. 10.3 we are using a micro switch with a roller; you could use the button or lever version. As you can see, the roller is fully extended when the tie bar moves across to change the points – this will push the roller bar towards the switch, changing the state of the switch. Fixing should be done by placing the micro switch as shown and then screwing it down with the right-hand fixing. Check it works before screwing the left-hand fixing.

In the circuit shown in Fig. 10.4 we are using a magnetic reed switch. As the tie bar is in the straight-through position the reed switch needs to be away from the magnet, therefore the reed switch is normally closed, which puts the green indicator on. When the tie bar moves towards the reed switch to change the points to turnout, the magnet influences the reed switch, which changes to switch the red indication 'on'. (Before you fix the reed switch check the tie bar does not hit the switch when it comes down.)

In Fig. 10.5 we are using a Gaugemaster seep motor, which has a switch built into the printed circuit board that holds the two solenoids. As the actuator pin is pulled from one side to the other a metal wiper blade makes contact with exposed copper areas of the printed circuit. The wiring for the motor is A, C and B. the changeover switch for the LEDs is D, E and F.

The Brimal MR319 has incorporated the switch and LED on one terminal board, therefore requiring only three wires to the point motor. The momentary switch energizes the point motor and a latching relay. The bi-colour LED will then show green for straight through, and red for turnout. This indication will be correct the next time you switch on your layout.

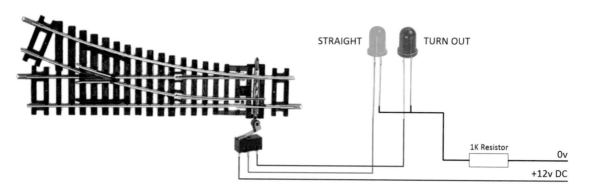

Fig. 10.3 Point status using a miniature microswitch with a roller actuator.

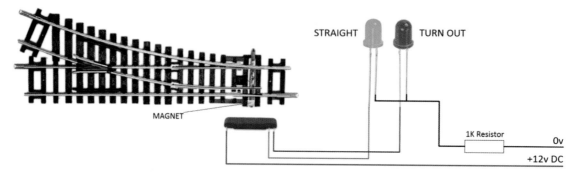

Fig. 10.4 Point status using a magnetic reed switch and magnets fitted to the actuator bar on points.

DETECTION | 125

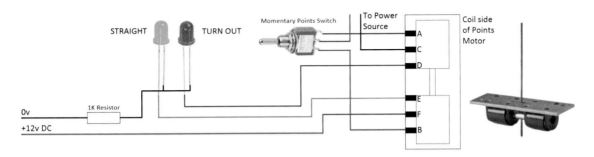

Fig. 10.5 Point status using the incorporated switch on the Gaugemaster seep motor. BRIMAL

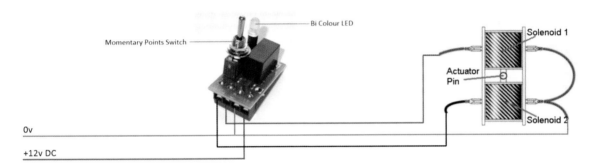

Fig. 10.6 Point status indication and actuation using the Brimal MR319.

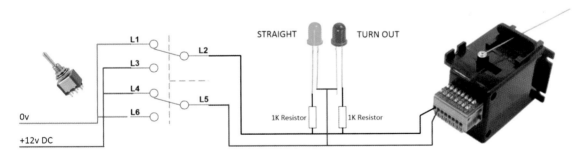

Fig. 10.7 Points indication using the Colbolt motor internal switch.

The cobalt motor shown in Fig. 10.7 is a stall-type motor so requires the input power to be reversed to change the state of the motor. To indicate the status, the LEDs need to be connected differently. The green LED is connected as shown, with the long lead through a resistor to the blue wire and the short lead to the black wire. The way the red LED is connected is the complete opposite to this.

The switch used is a double pole double throw, and is wired as shown. (Note the keyway is to the top when looking from the terminal end of the switch.)

All the above indicators except Figs 10.6 and 10.7 can be wired using wheat-germ bulbs or any type of miniature bulb rated at 12v. In all cases omit the 1K resistor.

THREE-WAY SLIP POINTS

The Peco SL-99 has three slip outlets, so it is useful to know which one is active at any one time. The circuit in Fig. 10.9 shows how you can wire three LEDs or three Lilliput lamps to the point motors. The indication could be either next to the points or back at the control panel. We are using the Peco PM1 points motor as it has a built-in changeover contact. It is not important where the actuator pin is while wiring up the contacts as long as you follow the drawing in Fig. 10.9. If you are using LEDs you will need to incorporate a dropping resistor of 1K in the negative wire. If you are using 12v lamps then the LED is not required.

TRAIN POSITION DETECTION

There will be times when you need to know where a train is if you have a large layout or you have hidden areas such as tunnels or behind scene track. There are two types of detection you may want to consider: momentary detection, or latching detection.

Let's look first at the types of detectors: these could be either magnetic switches or infrared switches. The magnetic switch consists of a magnet fixed to the undercarriage between the front wheels of the locomotive, and a reed switch positioned between the tracks.

The reed switch can be a small rectangular switch as shown in Fig. 10.11, or a round 6mm-diameter case fed in from the underside of the baseboard, as shown in Fig. 10.13. The distance between the magnet and sensor needs to be between 1 and 5mm. In Fig. 10.13 the sensor and magnet distance will need to be a lot closer, as the sensitive part of this sensor is half way down the casing.

Magnetic sensors normally only require a simple relay circuit; a few are described below.

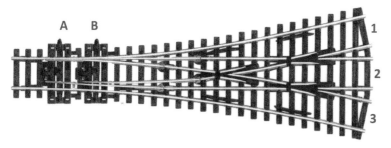

Fig. 10.8 A three-way slip point set. PECO

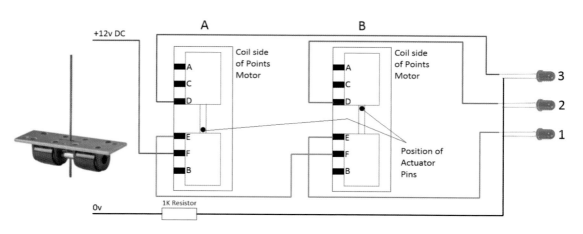

Fig. 10.9 Circuit diagram for a three-way slip using Gaugemaster seep motors.

DETECTION | 127

Fig. 10.10 Fitting a miniature magnet to the underside of a locomotive.

Fig. 10.12 Cross-section showing the installation of a tube-type reed switch.

TRAIN POSITION

The circuit shown in Fig. 10.13 is very basic. The principle is that the magnet on the loco switches the reed to 'on', this switches the LED on, and charges the capacitor. When the train continues past the reed, the reed switches off, but the LED remains on until the capacitor discharges through the resistor and LED. This could be approximately 30 seconds.

SIDING OCCUPIED

In this circuit the indicator on the control panel will start as green, indicating the siding is empty. The train moves into the siding, and before the buffer it

Fig. 10.11 Magnet on a locomotive and a reed switch between the sleepers on the track.

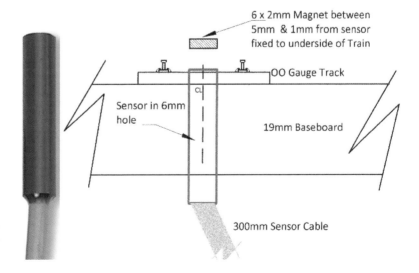

Fig. 10.13 Train position indicator placed where you require indication of a locomotive presence.

6 x 2mm Magnet between 5mm & 1mm from sensor fixed to underside of Train

OO Gauge Track

Sensor in 6mm hole

19mm Baseboard

300mm Sensor Cable

DETECTION

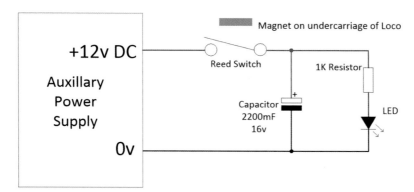

Fig. 10.14 Circuit diagram of a 'siding occupied' indication using a latching relay.

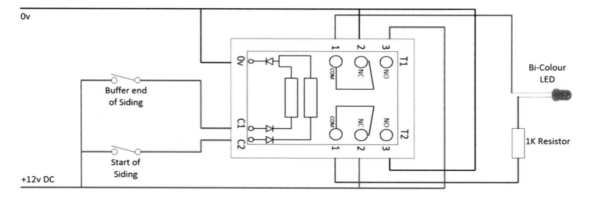

Fig. 10.15 Automatic reversing loop circuit diagram.

passes over a sensor changing the green to red, indicating the siding is occupied. When the train leaves the siding it passes over another sensor near the start of the siding, changing the red to green.

AUTOMATIC REVERSING LOOP

As explained in Chapter 7, Track Wiring, the polarity conflicts when it gets to IP15. The red rail which was on the bottom is now on the top and is joining to the blue rail. This cannot happen, so we install isolation fishplates at IP13 and IP15. This effectively isolates the straight section of track from the loop. As the train comes around the loop we need to change the polarity of the straight section to match what is on the right side of IP15. In Chapter 7, Track Wiring, we have shown two manual ways of doing this – now we are introducing a reed switch so it is done automatically. There are two operations to complete: first, change the points, and second, reverse the polarity in section C. The circuit shown in Fig. 10.16 achieves both these operations using the magnetic sensor SR1. If you have a loop A at the other end of the straight track you will have to repeat the circuit for that loop.

In this circuit we are using two latching relay boards, one to control the polarity of the track, and one to control the polarity of the stall-type motor. Although the wiring looks the same, which it is, they have to be different as the feed to the track is regulated by the speed controller, and the other feed is a permanent 12v DC supply. You need to synchronize both relays, as they may not be when received from the manufacturer. The best way to do this is to wire everything up as the circuit.

Now check the voltage on pin 1 of relay 1: it should be +12v DC. Do the same for relay 2: it should also be +12v DC. If relay 1 is correct and

DETECTION

Fig. 10.16 Layout of one side of a reversing loop.

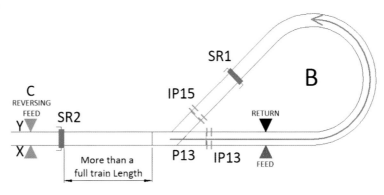

relay 2 is not, then put a +12v pulse on relay 2 C1: this will synchronize the two relays.

AUTOMATIC POINTS CHANGE

Part of the circuit shown in Fig. 10.17 can be used for automatically changing points. You may have a train in a siding or passing loop and you want to bring it out and then reset the points to straight through. Here, train 3 will move over sensor SR3, which will change points P5 to turnout, to allow the train to proceed out of the siding. SR4 must be far enough away from the points to allow the full length of the train to move past the points before SR4 is energized. This will set the points back to straight through. The circuit is shown in Fig. 10.19.

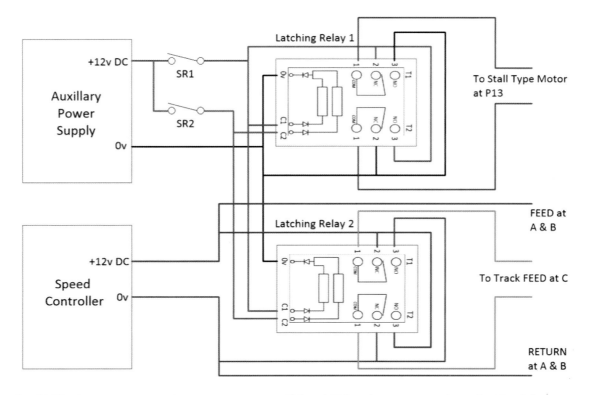

Fig. 10.17 Automatic points changing using sensors SR1 and SR2 and a magnet on the underside of the locomotive.

130 DETECTION

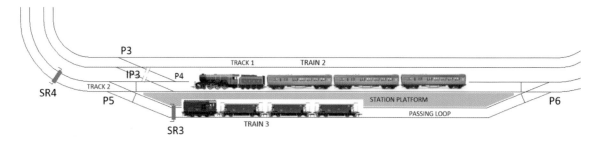

Fig. 10.18   Layout showing a passing loop.

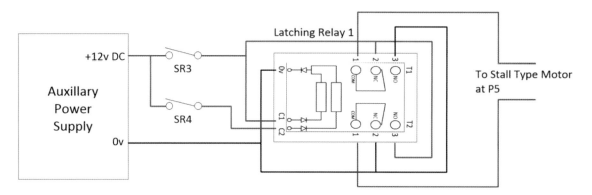

Fig. 10.19   Circuit diagram of the automatic points changer using a latching relay.

INFRARED DETECTION

Infrared radiation (IR) is electromagnetic radiation (EMR) with longer wavelengths than those of visible light, and is therefore generally invisible to the human eye. The unit requires a transmitter and a receiver which cannot see each other: when something reflective passes over them, the infrared beam is reflected to the receiver, switching on a relay or indicator lamp (see Fig. 10.21). The opposite is also possible where the receiver sees the transmitter all the time and only switches a relay or lamp when the beam is broken (see Fig. 10.22). Most circuits will have a timer circuit built into the unit so that you can adjust the 'on' time.

Fig. 10.20   Infrared sensor positioned between the sleepers. In this case the sleeper needs to be cut to incorporate the sensor.

DETECTION

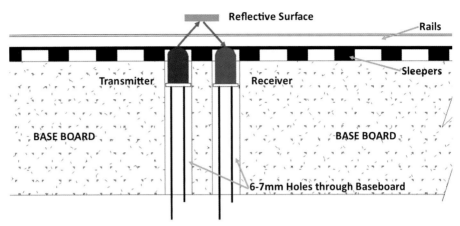

Fig. 10.21 An alternative infrared sensor using separate transmitter and receiver.

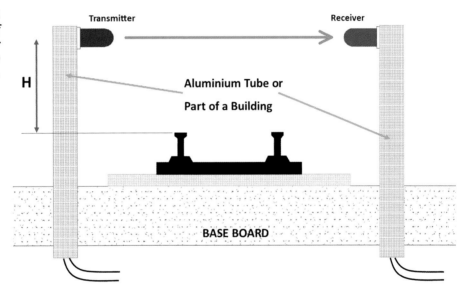

Fig. 10.22 A further example of separate transmitter and receiver for an infrared detector.

The sensor in Fig. 10.20 shows a transmitter and receiver in a single package with a divider between the two, the transmitter and receiver are available as individual units so could be placed as shown in Figs 10.21 and 10.22. The infrared sensor is placed between the rails as shown in Fig. 10.20 with the electronic circuit under the baseboard. For it to work well it requires a reflective surface on the underside of the loco. This can be either a little white paint, or the magnet that you are using for other jobs (as seen in Fig. 10.10). Again, if you are using magnets, check the clearance: you may have to set the infrared sensor much lower, or even on the underside of the baseboard with just a hole coming through to

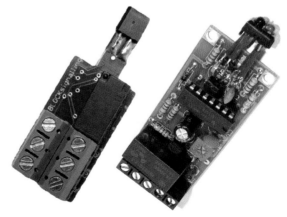

Fig. 10.23 Examples of infrared control circuits and detectors.

the track – as long as the sensor can see the loco, it will trigger. Depending on the sensor and the circuit, infrared does have a long detection range.

As with other forms of detection, once you have a signal it is down to you what you do with it. It could be used to detect the position of a train, switch sets of points, count carriages in a train, change signals, show positions on the control panel – the list is endless.

MIMIC DISPLAYS

Once you have your layout complete and your control panel has a diagram of your layout, you can install small LEDs to show the route of the train. This is called a mimic panel.

We will start with our final layout, as shown in Fig. 10.24.

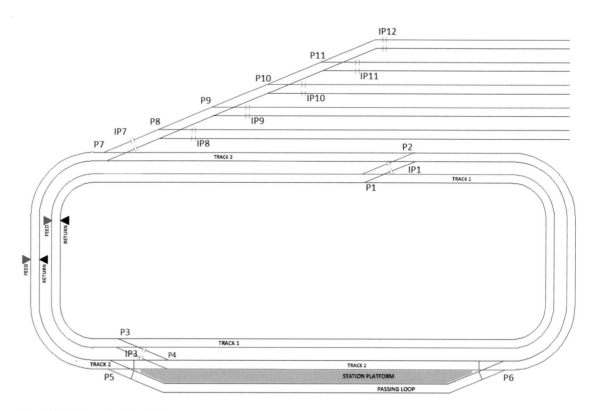

Fig. 10.24 *A typical simple layout.*

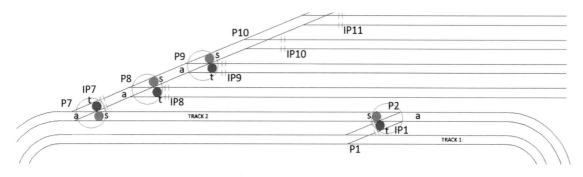

Fig. 10.25 *The fiddle-yard section of our typical layout with 3mm LEDs in the positions shown.*

DETECTION 133

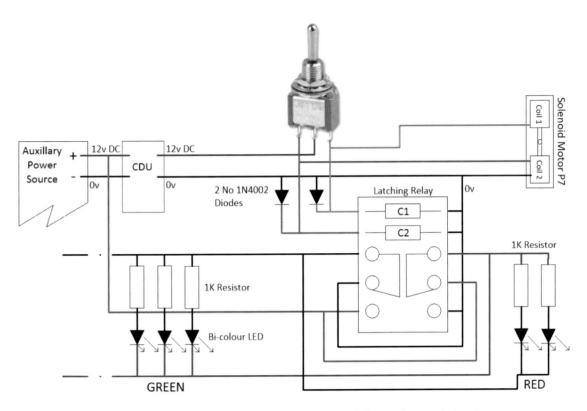

Fig. 10.26 Circuit diagram showing the points indication LEDs and the track route indication.

We will look at points P2, 7, 8 and 9 – with all the points in straight position the mimic panel would look as follows: track 2 would be green all the way to P2 and beyond; P7 TU shows a red LED; P8 is set at ST, so shows a green, as does P9. All the LEDs are Kingbright L-937EGW 3mm red/green LED bi-colour, so depending on the polarity will show either red or green. To do this we must add a relay to the points motor switch.

The circuit below starts with Fig. 38 from Chapter 7, Track Wiring. All we have done is add a second LED to the right contacts on the latching relay. On the left are the straight through LEDs, on the right are the red LEDs – you will need just the one. When you change the points P7 to TU all the colours connected to P7 will change, as shown in Fig. 10.27. Each set of points will be the same as this circuit.

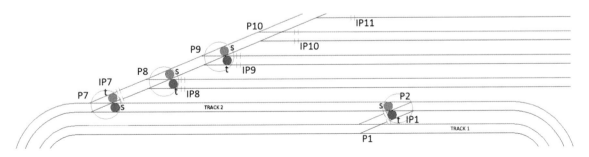

Fig. 10.27 The fiddle yard with the route from P7 to P9 in green.

134 | DETECTION

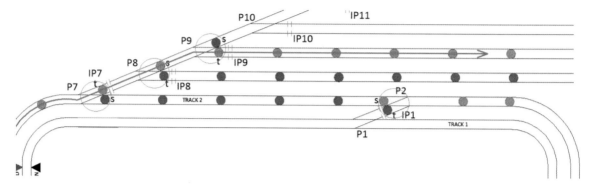

Fig. 10.28 The fiddle yard showing the route from P7 into siding at P9 all shown in green LEDs.

Now we can see that P7 has changed, and the ST LEDs have gone red and the TU LEDs have gone green. You will note that P2 has not changed so is still showing straight through; also P8 and P9 have not changed so the travel can only be up the spur to all the sidings. Changing either P8 or P9 will change that siding to green, giving a track from the main line to the changed siding.

This system of mimic display can be expanded to show the full track that you are using in green. The next image shows the train going from track 2 to siding 9. Here we have added more LEDs to the left side of the latching relay in the circuit above, to show the path the train will make. You would need to do this for each point set, so the others show red.

An alternative to this if you have a large fiddle yard is to use the LEDs in the siding to indicate which sidings are occupied and which are not. To do this you would use the circuit in Fig. 10.14, and add more LEDs in parallel to the existing one.

Note: When adding LEDs in parallel ensure that there is a resistor for each LED and not just one LED, as shown in Fig. 10.26. One LED will cause a voltage drop, which will reduce the brightness of the LEDs.

See Chapter 9, Lighting Projects, to see how an LED should be wired.

CHAPTER ELEVEN

TESTING AND TROUBLESHOOTING

MULTIMETERS

A multimeter is an essential tool when wiring a model railway system. There are many models available for sale, ranging from £5 to £500; however, I suggest you spend no more than £25 if you are only going to use it on your layout or around the house. Multimeters can do many tests, but there are four that you will use the most, so we will take a little time to familiarize you with them.

The layout will differ from one meter to the next, but the basic areas will be the same.

AC VOLTAGE

AC (alternating current) voltage is usually depicted by the symbol shown in Fig. 11.1a. Alternating current is normally the type of electricity in your household plug, and if you are running a DCC system it is the type of power being fed to the rails. On a basic DC layout you should not have AC or DCC going to the rails, but you may have AC going to some of the ancillary equipment.

Fig. 11.1a The standard sign on a multimeter for AC voltage.

DC VOLTAGE

DC (direct current) voltage is usually depicted by the symbol shown in Fig. 11.1b. Direct current is the power from batteries, and regulated power supplies such as battery chargers, phone chargers and universal power supplies. First set the function selector to the voltage next above 12v (this may be 50v, 100v or 200v). In a DC system the tracks are powered by

Fig. 11.1b The standard sign on a multimeter for DC voltage.

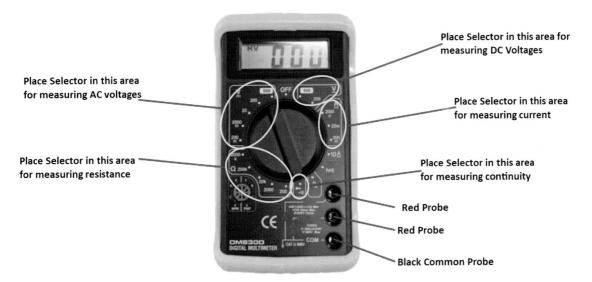

Fig. 11.1 A typical multimeter.

TESTING AND TROUBLESHOOTING

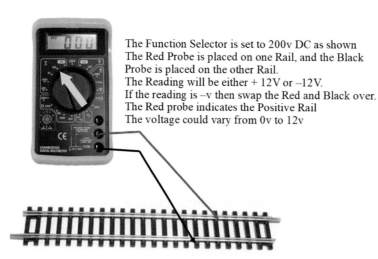

The Function Selector is set to 200v DC as shown
The Red Probe is placed on one Rail, and the Black Probe is placed on the other Rail.
The Reading will be either + 12V or –12V.
If the reading is –v then swap the Red and Black over.
The Red probe indicates the Positive Rail
The voltage could vary from 0v to 12v

Fig. 11.2 Using the multimeter to check voltage.

a variable 12v DC supply from the controller, so by placing the black probe on one rail and the red probe on the other rail you should get a reading between 1v and 12v.

CURRENT (AMPS)

Amps (or current) is usually depicted by an 'A'. The multimeter is used to measure the current you are using. As can be seen from the drawing, the meter probes are placed between the feed wire and the rails. This means the power has to flow through the multimeter to get to the track. Here the meter measures the amount of current the train is using, and gives a reading.

CURRENT DRAWN BY A POINT MOTOR (SOLENOID)

It is sometimes useful to know what other pieces of equipment use, as there is a point where you may have too many items being supplied from one power supply and it can no longer cope. If this is the case a fuse will blow, or the power unit will overheat and eventually fail.

The most common items will be point motors and signals. Coil-type point motors have two coils, but you only need to measure one, as only one will ever be used at the same time. In this diagram, one wire to a point-motor coil has been removed, and the amp-meter has been placed in the circuit, so when

Speed Controller

One wire feed from the controller goes to the Track as shown. The other wire goes to the Red Probe on the Multimeter and the Black Probe goes to the track. As you run the Train the reading will show the current being drawn by the Train.

Fig. 11.3 Using the multimeter to monitor current.

TESTING AND TROUBLESHOOTING

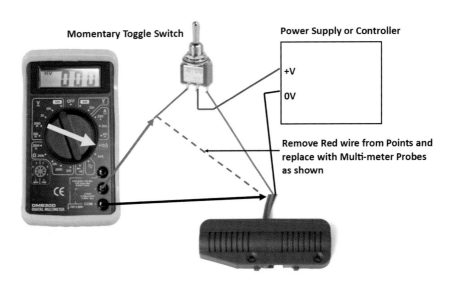

Fig. 11.4 The multimeter being used to check the current drawn by a typical point motor.

that coil is energized via the toggle switch you will be able to read the current being used. The reading should be in milli-amps.

CONTINUITY TESTER

This also seconds as a diode tester, of which you may have a few on your layout. As a continuity tester you need the red probe in the socket marked VΩμA: by joining the two probes together you should hear a sound – this indicates a continuous circuit through the probes.

In the diagram we are trying to find the break in the track which is stopping the train. One way is to see where the train stops, and that will be where the problem is. The other way is to use the continuity tester on the multimeter. Break the track at 'X', place the probes with the red one where shown, and

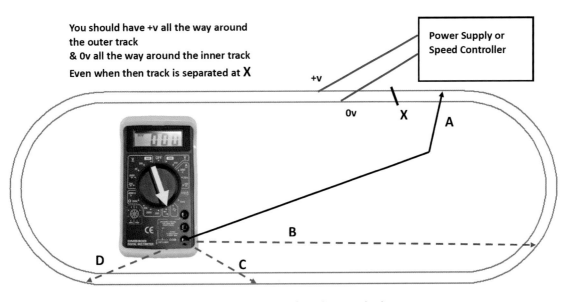

Fig. 11.5 Using the multimeter to check the continuity of track around a layout.

TESTING AND TROUBLESHOOTING

the black in position 'A'. If the track is all right, the buzzer will sound; if not, move the probe to position 'B' and so on, until the buzzer sounds. The problem will be somewhere between the last one and the previous one. Normally this will be a fishplate that is not connecting properly. This is just one example of continuity. The same principle can be used at points and crossovers.

Sometimes you don't want continuity (where a section of track has been isolated), and this meter can test for that as well.

BEFORE YOU APPLY POWER TO YOUR LAYOUT

An important continuity test is between the live and return tracks. This test should be done before you switch power on to the tracks. Put one probe on one rail and the other probe on the second rail – there should be no buzzer and no reading on the meter. If the buzzer sounds you have a short and could blow the power unit. The short needs to be found before you switch 'on'.

Clip the red probe to the live rail and the black probe to the return track. Now uncouple sections of track, starting furthest away from the meter connection; reconnect if there is no change and move on to the next section, until the buzzer stops. This will tell you where the problem is.

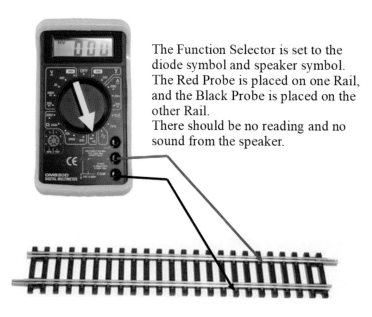

The Function Selector is set to the diode symbol and speaker symbol. The Red Probe is placed on one Rail, and the Black Probe is placed on the other Rail.
There should be no reading and no sound from the speaker.

Fig. 11.6 Using the multimeter to check you have no continuity between the live rail and the return rail.

APPENDIX I: GLOSSARY OF TERMS

alternating current/AC The electric current that flows alternately through each wire in a rapid sequence called 'a cycle'. The normal mains electricity supply is powered by alternating current/AC.

ampere or **amp** The current flowing in an electrical circuit.

armature The revolving part of an electric motor.

back to back The measurement between the backs of the running wheels on rolling stock.

ballast Usually stone chippings, or similar material, that supports the sleepers and holds them in place on the ground.

baseboard The structure on which to build a model railway layout.

bogie An assembly that supports the wheels on a coach, freight wagon or locomotive. Each bogie usually holds two or three pairs of wheels.

brushes Pieces of carbon fitted into a holder on either side of the commutator on an electric motor. These are often copper or phosphor-bronze strips that touch either side of the commutator.

buffer beam The traverse beam at each end of a locomotive, coach or wagon which carries the buffers. Sometimes called a headstock.

catch point A single trailing point blade set into an ascending track to derail wagons that have come uncoupled and are running back down a hill. A catch point prevents runaways colliding with a following train.

CDU (capacitor discharge unit) Used to store power so that several point motors may be operated simultaneously.

catenary Overhead wires and their supports for carrying electricity above the railway.

chair Metal clips that hold the rail and are bolted or spiked to the sleepers.

check-rail A rail inside the running rails seen on curves and points to keep the flanges of the wheels from riding up on the opposite rail.

colour light signal A signal that uses only coloured lights to provide an indication of how far in front of the train the line is clear.

common return A wire connecting one side of the running rail which returns the current from all round the track or layout.

commutator The slotted copper segments at the end of the armature on an electric motor, which transfers the current from the brushes to the coils wound on the armature.

connecting rod The rod on locomotives that connects the piston rod (on the crosshead) to the driving wheel. The little end of the connecting rod is at the piston rod end and the big end is at the driving wheel end.

controller An instrument controlling the speed and direction of a model locomotive by means of a variable resistance, a variable transformer, or by an electronic circuit.

coupling rod The rod connecting the large wheels of a locomotive.

crank The pivot point or pin where the connecting rod joins the driving wheels.

crosshead The two parallel pieces of metal connecting the piston rod which slide in the side bar and transmit power to the connecting rod.

crossing The diamond formed by two tracks where they cross one another.

crossover A crossing from one set of rails to another.

cutting A section of railway line where the surrounding countryside is at a higher level than the line and the ground has been dug away to put in the line.

cut-out A safety switch that cuts out the electrical supply in the event of a short circuit or overload. Modern cut-out switches can often be reset by pressing a button.

DCC (digital command and control) The application of computer technology to control the movements of locomotives. Each locomotive is fitted with a decoder (or 'chip'), which is uniquely programmed and recognizes its own identity, and responds only to those control signals that are addressed to it. DCC also allows a wide range of extras including controllable lighting and on-board sound.

diamond The centre portion of an acute angled crossing.

DC (direct current) The opposite of 'alternating current'/AC. DC is a current that flows in one direction. Most model railway locomotives work on a direct current of 12v.

disc wheels Wheels that are solid, *ie* have no spokes. Sometimes a disc wheel will have holes around its circumference.

distant signal A semaphore signal giving the driver advanced warning of the position of the next home signal, allowing him to slow the train if it were at danger. The facing arm of the distant signal is yellow with a fishtail.

double crossover Two crossovers superimposed on one another, *ie* in a facing and trailing direction.

embankment A section of railway line where the surrounding countryside is at a lower level than the line and the ground has been built up to put in the line.

facing point A turnout or point that faces the oncoming trains.

fiddle yard Used on model layouts to store complete trains which can be ready to run out of the fiddle yard on to the main circuit of the layout. It can also be used for changing trains.

fine scale A smaller scale nearer to the true scale of the prototype; for example fine scale 'OO' gauge is often described as EM gauge.

fishplate A metal plate which clamps on the end of two abutting rails to make sure the rail is in line. On a model railway a metal fishplate will conduct electricity across the gap. If an isolated section is required then a plastic fishplate is used.

flyover A bridge carrying railway tracks over another railway line instead of having them cross on the same level.

footplate The cab floor of a locomotive, or the plate of the platform running along each side of a locomotive's boiler.

four-aspect signal A colour light signal using four lights. From top to bottom in order, the lights are yellow, green, yellow, red. A red indication means the next section contains a train. A single yellow light (using the lower yellow aspect) means the next section is clear but there is a train in the following section. A double yellow means the next two sections are clear but there is a train in the third section. A green indicates that the next three sections are clear.

gantry Bridge or walkway over railway tracks on which several signals are mounted.

gauge The distance between running rails of the railway track – 4ft 8½in (1m 43cm) in England.

gears Large and small toothed or cog wheels that mesh with each other to give an increase or decrease in speed.

ground frame Small signal box or lever frame often seen in shunting yards. It does not usually control signals or points for a main line but for sidings and yards.

handrail A rod or tube along the boiler of a locomotive to act as handgrips for the loco crew. They can also be found on cab sides and brake vans to assist the passenger in getting on and off the vehicle.

home signal The signal that controls entry into a section. It either shows the line is clear (the signal arm would be up) or that another train is ahead (the signal arm would be horizontal).

inner home signal A home signal within station limits where an outer home signal is positioned. The inner home signal is in advance of the outer home and usually to the rear of the signal box.

jinty The nickname for a six-wheel side-tank locomotive designed by Sir Henry Fowler for shunting and light freight duties.

junction signal Any signal that has more than one route and is capable of displaying an indication of which route has been selected. A junction indicator will also be fitted to a junction signal to inform the driver which way a junction is set, by means of white lights.

kettle A lineside boiler for filling locomotive boilers directly with hot water.

king lever A lever in a signal box which cuts out the box's control and allows its signals and points to be controlled remotely from another box, or automatically via track circuits.

kip A hump at the top of a rope-hauled railway to prevent wagons or carriages accidentally running back down the incline; an incline on which wagons are built to be run off by gravity as required, usually at a colliery, to feed a loading point.

lever frame The assembly that holds the signal and point levers in a signal box or ground frame. A lever frame is made up of slots for the levers to operate in, and allows for them to be locked together.

live steam A method of powering a locomotive, as opposed to electric current. Hornby was the first manufacturer to produce commercially a live steam locomotive, powered entirely by steam, in 'OO' gauge.

loading gauge The size limit for locomotives, freight and passenger rolling stock over a specific stretch of railway. This is measured horizontally in relation to the position of platform edges and tunnel walls, for example, and vertically in relation to bridge and tunnel heights. The loading gauge can vary considerably for the same track gauge, especially in other parts of Europe and North America. It is a metal frame, often found in freight yards, suspended over the track to indicate the limit an open freight wagon may be safely loaded.

lower quadrant signal A semaphore signal that lowers its arm to indicate that the line ahead is clear.

louvres Horizontal slots cut into the sides of diesel and electric locomotives and certain goods vans to give ventilation.

main frame The strong metal sides of the chassis of a locomotive in which holes are drilled for axles. They also carry the spacing pieces and fixings for the locomotive's motor.

multiple unit A set of coaches that are self-powered and have a driving compartment at each end. Multiple units are usually powered by an electric motor or a diesel engine.

narrow gauge A railway running on tracks that have a closer distance between the rails than normal.

NEM coupling A standard universal type of coupling that can be fitted by means of a small socket on the underside of most locomotives and rolling stock. It enables models from different manufacturers to be run coupled together.

ohm The measurement of the resistance in an electrical circuit.

outer home signal An additional signal placed before the home signal that protects trains shunting back past the home signal. It also indicates station limits at the approach end of a station.

overload Where the electrical load (*ie* several model locomotives running at the same time) requires more power than the transformer or controller can give.

pantograph A metal assembly on the roof of an electric unit to collect current from an overhead wire.

point One railway track turning into two or three tracks, or the crossing of one track with another.

point motor An electric motor or solenoid used for changing the points.

point rodding The rodding or wires that move the point either from an electric motor or solenoid or from the lever in a signal box.

power unit Transformer and rectifier used to convert mains electricity (normally 240v AC) to the smaller voltage required by a model railway controller (normally 12v DC) or digital command and control systems (DCC).

prototype The full size locomotive, coach or wagon; full size railway practice.

push/pull A type of train where the carriages are kept permanently coupled to the locomotive, which pulls them in one direction then pushes them in the other.

GLOSSARY OF TERMS

rain strips Curved pieces of wood or metal fixed on a coach roof to prevent the rain running down the sides when the doors are opened.

ramp The sloping end of a station platform; sloping object in the centre of the running rails in model railways usually used for uncoupling.

ready to run A model which can be taken straight out of the box, placed on your layout, and run. All Hornby locomotives and rolling stock are manufactured as 'ready to run'.

rectifier An electrical item used for changing alternating current (AC) to direct current (DC).

relay An electrical device for switching currents to other circuits; the opening or closing of a circuit. A relay can also be used in place of a point motor.

resistance A measurement of electricity; a substance that reduces the flow of electricity.

reverse loop A model railway track that loops 180° to turn trains round to the direction from which they came.

reversing switch Electrical switch that changes the polarity of the electrical supply to the model railway and thus reverses the direction of the locomotive.

rolling stock Anything with wheels on it which can run on the track, including locomotives, carriages, freight wagons and maintenance vehicles.

saddle The cradle in which the smoke-box end of the boiler of a locomotive rests.

scale The relationship in size between the model and the full size item.

scissors crossover The facing and trailing crossovers between two adjacent tracks.

section In railway terms a length of track, usually between two signals.

semaphore signal A signal that uses a moving arm to indicate the state of the line ahead. Having the signal arm pointed horizontally usually indicates that the next section contains a train. A raised or lowered arm indicates that the line is clear. The arm also has coloured filters fitted, which are moved in front of a lamp when the signal arm moves to help drivers see the signal at night.

short circuit The negative and positive wires of an electrical supply touching one another. One example of a short circuit is when a metal object, such as a screwdriver, is placed on the track providing a path for electricity from one electrical circuit to another.

shunting The movement required to rearrange the position of wagons or coaches in a train; to pick up and set down wagons in a goods train.

shunting signal A smaller than normal signal that is used specifically to indicate whether or not particular shunting moves may take place. A shunting signal is usually represented on a modern image layouts with a colour light ground signal.

signal box A building from which the surrounding points and signals are operated. May contain either a lever frame, or in more modern signal boxes a panel containing switches and coloured lights.

six-foot way The distance between two railway lines on a railway.

sleeper The wooden or concrete beam on which the rails rest and are kept in position by a chair.

slide bar The two parallel metal bars or strips in which the crosshead slides, forming part of the valve gear.

solenoid The effect of a current passed round the coil of a solenoid producing a forced magnetism that pulls down the solenoid's centre core, providing energy to operate points and signals. A solenoid is similar in action to a relay.

spectacle plate Windows at the cab front enabling the driver and fireman to have forward vision.

splashers The coverings on the upper part of a footplate protecting the driver and fireman from being splashed by rain or mud.

spur drive The drive through a chain of gears.

spring motor A clockwork mechanism.

starting signal An extra signal placed at the departure end of a platform to allow trains to run into the station and stop, even though there is a train in the next section.

three-aspect signal A colour light signal using three lights. From top to bottom, in order, the lights are green, yellow and red. A red light means the next section contains a train. A yellow light means the next section is clear but there is a train in the following one. A green indicates the next two sections are clear.

TOPS (total operations processing system) A computer-based program developed by British Rail to monitor the movements of all freight and passenger rolling stock, and locomotives. The system was introduced in the early 1970s to record every movement of freight traffic and non-multiple-unit passenger trains. The information is broadcast live to a central computer to form a comprehensive and up-to-the-minute picture of the freight and passenger traffic situation over the whole of the rail network.

trailing points Turnouts (points) or crossovers that are against the direction of travel, *ie* a train has to reverse to pass over them.

two-aspect signal A colour light signal using two lights. A two-aspect signal can either be yellow and green, or red and green. As with other colour-light signal formations, a yellow aspect means the next section is clear but there is a train in the following one. A green indicates the next two sections are clear.

turnout *See* points.

uncoupling ramp A ramp fitted under or between the rails on a model railway to uncouple rolling stock remotely. It is not a prototypical example, but uncoupling ramps can often be seen in hump shunting yards.

underframe The chassis of a wagon or coach.

upper quadrant signal A semaphore signal that raises its arms to indicate that the line ahead is clear.

valve gear The mechanism used to power the driving wheels and pistons of a locomotive. In most steam locomotives the valve gear is exposed, although in Bulleid's West Country Class, for example, the chain-driven valve gear is enclosed within the body.

vestibule The corridor connection between coaches, usually a flexible gangway to enable passengers to pass between coaches in a train.

volt An electrical measurement meaning the pressure of electricity in the supply.

watt A unit of electrical measurement to describe the energy produced. A watt is the number of volts multiplied by the number of amps.

wheel arrangements The arrangement of wheels on a locomotive. There are many varieties of wheel arrangements for steam and diesel locomotives, depending on the number of leading, driving, trailing or load-bearing wheels there are.

'Y' point A turnout in the shape of a letter 'Y'; a single track turning into two parallel tracks.

APPENDIX II: ELECTRONIC SYMBOLS

+ −		Battery
		Bell
AC / −DC / +DC / AC		Bridge rectifier
		Buzzer
+		Capacitor electrolytic
		Capacitor non polarized

ELECTRONIC SYMBOLS

		Circuit breaker
		Coaxial cable (or shielded cable)
		Coil
		Conductors joined
		Conductors unjoined
		Diode
		Filament lamp
		Fuse

ELECTRONIC SYMBOLS

(ground to chassis symbol)		Ground to chassis
(ground to earth symbol)	EARTH / LIVE / NEUTRAL	Ground to earth
—(A)—		Meter amps
—(V)—	DC VOLTS	Meter volts
—(M)—		Motor
ANODE —▷|— CATHODE, GATE		Thyristor

ELECTRONIC SYMBOLS

ANODE ▶|– CATHODE		Diode zener
ANODE ▶|– CATHODE		Light-emitting diode (LED)
L1 –o o– L2		Relay coil
		Relay with twin coils
–/\/\/\–		Resistor
		Resistor light sensitive
		Resistor preset
		Resistor variable

ELECTRONIC SYMBOLS

		Speaker
		Switch single pole single throw (SPST)
		Switch double pole single throw (DPST)
		Switch single pole double throw (SPCO)
		Switch double pole double throw (DPDT)
		Switch triple pole double throw (3PDT)

ELECTRONIC SYMBOLS 149

L1, L2, COM schematic	(toggle switch photo)	Switch (SPDT) centre biased
L1, L2, L3, COM schematic	(toggle switch photo)	Switch single pole triple throw (SP3T)
L1, L2 pushbutton schematic	(push button photo)	Switch push button normally open
L1, L2 pushbutton schematic	(push button photo)	Switch push button normally closed
Rotary switch schematic (positions 1–12)	(rotary switch photo)	Switch rotary

ELECTRONIC SYMBOLS

		Switch thermal (circuit breaker) normally closed
		Switch thermal (circuit breaker) normally open
		Switch magnetic reed (glass)
		Switch magnetic reed (encapsulated) changeover
		Switch magnetic latching reed
		Terminal
		Thermistor (temperature sensitive resistor)
		Thyristor

ELECTRONIC SYMBOLS 151

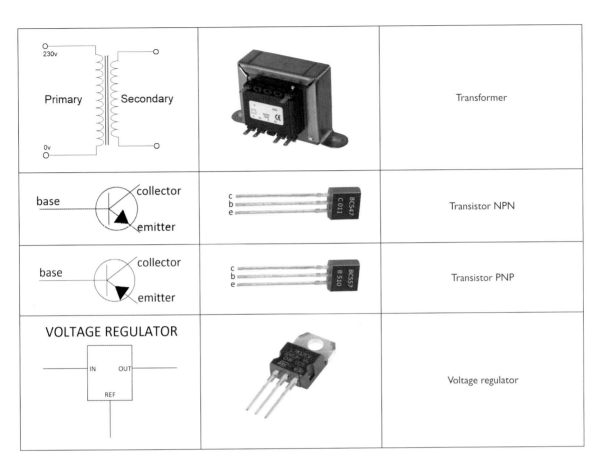

Transformer symbol (Primary 230v, Secondary 0v)	Transformer	Transformer
Transistor symbol (base, collector, emitter)	BC547 C011	Transistor NPN
Transistor symbol (base, collector, emitter)	BC557 B510	Transistor PNP
VOLTAGE REGULATOR (IN, OUT, REF)		Voltage regulator

APPENDIX III: MODEL RAILWAY TRACK SYMBOLS

You will find hundreds of layout drawings on the internet and in books, and it is useful to understand what all the symbols mean on the drawing. Below we have a basic layout on which we have used most of the symbols you will come across. The table gives an explanation of each of these symbols.

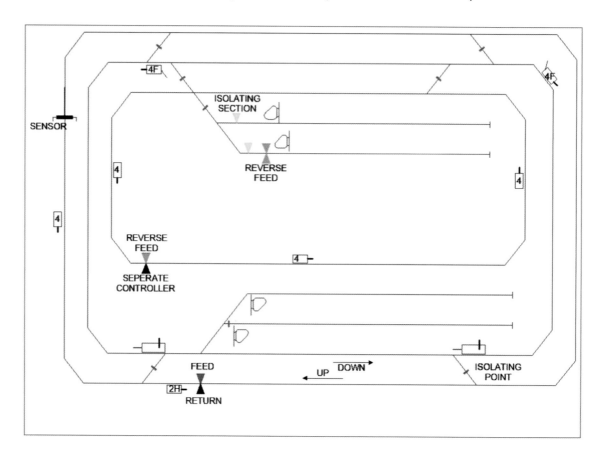

FEED / RETURN	The single black line is in fact showing the two-rail track. So the red arrow is pointing to the top rail and is identifying this rail as the *feed* (positive voltage to the rail). The black arrow is pointing to the lower (bottom) rail and is identifying it at the *return* (negative or 0v rail)
REVERSE FEED / SEPERATE CONTROLLER	This symbol is showing that the track has a reversible feed and is powered by a separate controller. Commonly used in a siding or fiddle yard.

MODEL RAILWAY TRACK SYMBOLS

REVERSE FEED	This symbol is showing that this section of track is going through a reversing switch so, depending on the position of the switch, the top rail could be either *feed* or *return*; the same applies to the lower rail.
ISOLATING SECTION	This shows that this section of track can be isolated from the main layout. This is normally by way of a toggle switch.
	This shows where a point set is used to isolate one track from another. This could be a single point set or dual points, or a crossover.
	A sensor could be anything, such as a microswitch, reed switch, leaf switch or infrared switch. It is placed wherever you want something to happen, such as to indicate a locomotive position, or use the loco to change something on the layout, such as signals or points, or sound.
UP	The normal direction of the train with all switches and controllers in their normal state.
DOWN	The normal direction of the train with all switches and controllers in their normal state.
REVERSIBLE	This section of track is reversible.
	Lamp signal with the number of aspects written in the box, with home or distant.
	Lamp signal with feather. Could be two-, three- or four-aspect.
	Ground signal.
	Semaphore signal.

APPENDIX IV: CAPACITOR CONVERSION CHART

uF / MFD	nF	pF / MMFD	uF / /MFD	nF	pF / MMFD
1,000	1,000,000	1,000,000,000	6.8	6,800	6,800,000
680	680,000	680,000,000	5.6	5,600	5,600,000
470	470,000	470,000,000	5.0	5,000	5,000,000
240	240,000	240,000,000	4.7	4,700	4,700,000
220	220,000	220,000,000	4.0	4,000	4,000,000
150	150,000	150,000,000	3.9	3,900	3,900,000
100	100,000	100,000,000	3.3	3,300	3,300,000
88	88,000	88,000,000	3.0	3,000	3,000,000
85	85,000	85,000,000	2.7	2,700	2,700,000
82	82,000	82,000,000	2.2	2,200	2,200,000
80	80,000	80,000,000	2.0	2,000	2,000,000
75	75,000	75000,000	1.8	1,800	1,800,000
72	72,000	72,000,000	1.2	1,200	1,200,000
70	70,000	70,000,000	1.0	1,000	1,000,000
68	68,000	68,000,000	0.82	820	820,000
65	65,000	65,000,000	0.68	680	680,000
64	64,000	64,000,000	0.47	470	470,000
60	60,000	60,000,000	0.33	330	330,000
56	56,000	56,000,000	0.22	220	220,000
53	53,000	53,000,000	0.20	200	200,000
50	50,000	50,000,000	0.10	100	100,000
47	47,000	47000000	0.01	10	10,000
45	45,000	45000000	0.0068	6.8	6,800
43	43,000	43000000	0.0047	4.7	4,700
40	40,000	40000000	0.0033	3.3	3,300
39	39,000	39,000,000	0.0022	2.2	2,200
36	36,000	36,000,000	0.0015	1.5	1,500
35	35,000	35,000,000	0.001	1	1,000
33	33,000	30,000,000	0.00068	0.68	680

CAPACITOR CONVERSION CHART

uF / MFD	nF	pF / MMFD	uF / /MFD	nF	pF / MMFD
30	30,000	30,000,000	0.00047	0.47	470
27.5	27,500	27,500,000	0.00033	0.33	330
27	27,000	27,000,000	0.00022	0.22	220
25	25,000	25,000,000	0.00015	0.15	150
24	24,000	24,000,000	0.0001	0.1	100
22	22,000	22,000,000	0.000068	0.068	68
21	21,000	21,000,000	0.000047	0.047	47
20	20,000	20,000,000	0.000033	0.033	33
19	19,000	19,000,000	0.000022	0.022	22
18	18,000	18,000,000	0.000015	0.015	15
16	16,000	16,000,000	0.00001	0.01	10
15	15,000	15,000,000	0.0000068	0.0068	6.8
12	12,000	12,000,000	0.0000047	0.0047	4.7
10	10,000	10,000,000	0.0000033	0.0033	3.3
8.2	8,200	8,200,000	0.0000022	0.0022	2.2
8.0	8,000	8,000,000	0.0000015	0.0015	1.5
7.5	7,500	7,500,000	0.000001	0.001	1

APPENDIX V: ENAMELLED COPPER WIRE

Magnet wire or enamelled wire is a copper or aluminium wire coated with a very thin layer of insulation. It is used in the construction of transformers, inductors, motors, speakers, hard disk head actuators, potentiometers, electromagnets, and other applications that require tight coils of wire. The wire itself is most often fully annealed, electrolytically refined copper. Aluminium magnet wire is sometimes used for large transformers and motors. An aluminium wire must have 1.6 times the cross-sectional area as a copper wire to achieve comparable DC resistance.

Due to this, copper magnet wires contribute to improving energy efficiency in equipment such as electric motors, solenoids.

Smaller diameter magnet wire usually has a round cross-section. This kind of wire is used for things such as electric guitar pick-ups.

The following is a conversion table: American wire gauge (AWG), standard wire gauge (SWG), and diameter in millimetres. (CSA – cross-sectional area.)

STANDARD TO AMERICAN WIRE GAUGE CONVERSION CHART

AWG	SWG	Diameter mm	CSA sq mm	Fusing current amps	Fuse rating amps	Wire rating amps
32		0.202	0.032	7	4	0.4
	35	0.214	0.036	8	4	0.5
31	34	0.226	0.04	9	5	0.5
	33	0.250	0.049	10	5	0.6
30		0.255	0.051	10	6	0.7
	32	0.269	0.057	11	6	0.8
29	31	0.288	0.065	12	7	0.9
	30	0.315	0.078	14	7	1.0
28		0.321	0.081	15	8	1.1
	29	0.331	0.086	15	9	1.1
27		0.362	0.103	17	10	1.4
	28	0.397	0.124	20	10	1.6
26		0.404	0.128	21	10	1.7
	27	0.410	0.132	21	12	1.7
25	26	0.454	0.162	24	14	2.1
	25	0.496	0.193	28	15	2.5
24		0.517	0.21	30	17	2.8
	24	0.559	0.245	33	17	3.2
23		0.574	0.259	35	19	3.4

ENAMELLED COPPER WIRE 157

AWG	SWG	Diameter mm	CSA sq mm	Fusing current amps	Fuse rating amps	Wire rating amps
	23	0.613	0.295	38	21	3.9
22		0.642	0.324	41	25	4.3
	22	0.723	0.41	49	25	5.4
21		0.724	0.412	49	29	5.4
	21	0.807	0.511	58	31	6.7
20		0.841	0.556	62	34	7.3
19		0.900	0.636	68	35	8.4
	20	0.917	0.66	70	41	8.7
	19	1.017	0.813	82	42	10
18		1.026	0.826	83	49	11
17		1.151	1.04	99	54	14
	18	1.221	1.17	108	60	15
16		1.306	1.34	119	69	18
	17	1.441	1.63	138	71	21
15		1.463	1.68	141	83	22
14	16	1.627	1.08	166	99	28
13		1.830	2.63	198	99	35
	15	1.833	2.64	199	116	35
	14	2.034	3.25	232	139	43
11		2.293	4.13	278	147	54
	13	2.386	4.47	295	167	59
10		2.588	5.26	333	170	69
	12	2.620	5.39	339	197	71
9	11	2.899	6.6	395	237	87
8		3.270	8.4	473	270	110
	10	3.568	10	539	262	130

INDEX

7/0.2 wire 53
16/0.2 wire 53
16v AC 46
24/0.2 wire 53
32/0.2 wire 53

actuator 84, 86
alternating current (AC) 135
aluminium cable clips 57, 83
ammeter 49
anode 31
anti-surge fuse 33
arc flash 117
automatic signals 113
automatic train detection system (AWS) 6

banking and tilt 15
baseboard 12
bimetal strip 34
breakdown voltage 35
bridge rectifier 31, 144
building lighting 119
busbar 83

cable clips 57
cable markers 58, 59
cable ties 57
candela 38, 115
Capacitor Discharge Unit (CDU) 87
capacitor (electrolytic) 36
capacitor equivalents 154
carriage lighting 119
cathode 31
centre biased 68, 85, 87, 93
centre OFF 68, 85, 87, 91
circuit breaker 34, 145
crossover 24
clamp meter 136
closure rail 18
coil-operated 43

colour codes 80
colour light signals 107
continuity links 20
continuity tester 137
control panels 60, 64
copper wire 156
crossover scissor 24
current drawn 136
cycles 32

dead area 18
decoder 9
desktop enclosures 64
diamond crossing 96
Digital Command Control (DCC) 9, 10
diode 31, 145
diode matrix 101
direct current (DC) 9, 10, 135
double slip 97
double pole (DP) 73, 78
double pole, double throw (DPDT) 73, 78
double throw (DT)

Electro Motive Force (EMF) 43
electromagnetic 42, 104
electrofrogs 18
electrolytic capacitor 35, 144
encapsulated switch 75
equipment wire 52

fiddle yard 26, 101, 133
filament lamp 37, 145
fillet 21
fishing 21
fishplate 13, 18
forward/reverse switch 85, 91
four pole, double throw (4 PDT) 79
frog 18
fuse 33, 145

gauge 7
glass reed switch 75
ground signals 111
guide rails 18

heatsink 42
Helices 15

infrared detection 130
insert control panels 60
insulated frog 18
isolation 22
isolation fishplates 22

junction block 56, 84

lamp 115
layout (basic) 16,
leaf switch 76
LED. holder 62
LED (Light Emitting Diode) 37, 115, 122, 147
level crossing 116
lever coupler
lighthouse beacon 117
loop (reversing) 26, 91, 93, 128
lumens 38

magnetic switch 75
making and breaking 66
mimic displays 132
microswitch 74
momentary switch 68
multicore cable 53
multimeter 49, 135
multi-strand wire 52

ohm 39
ON–OFF switch 69
ON–ON switch 70
ON–ON–ON switch 72
overhang 14

P clips 57, 58
parallel connection 30
passing contact switch 66

passing loop 22, 90, 130
plastic fishplates 22, 89
points 17
point motors 17, 84, 125, 129
points status memory 93, 123, 133
polarity switch 99
polarity reversing 85
positive supply 48
power connections 81
Power Supply Unit (PSU) 45
preset resistor 40
primary winding 30
printed circuit 50
push button switch 62, 68, 149

quick blow fuse 33
quick splice connector 83

radial 36
radius 15
railroad switch 17
ramps 15
rectifier 32
rectifier – bridge 32
reed switch 75, 124, 127, 150
regulator – voltage 41
relay 42, 147
Residual Current Device (RCD) 51
resistor 39, 147
resistor - preset 40, 147
resize scale 9
reversing loop 26, 91, 93, 128
roller lever switch 74
rotary switch 62, 72, 149
rotational motion 45

scale 7, 12,
scissor crossover 97
screened cable 54
secondary winding 29, 30
semiconductor 35, 37
semaphore signals 106
series connection 29
shorting clips 20
shunting yard 132, 61

shuttle 94
sidings 23, 127
signal diode 31
signals at points 110
simple layout 16
single pole, double throw (SPDT) 78, 148
single pole, single throw (SPST) 78, 148
sleeper
slide switch 71
slip (double) 97
slip (triple) 98
slow blow 33
slow-motion motors 86
solder tags 57, 150
solenoid 84
speed controller 47
spring lever switch 76
staging yard 27
stall motors 86
star wheel 76
station stop 96
stock rail 18,
stranded wire 52
street lights 118
super-elevation 15
switch 43, 62, 109

terminal block 56
three-aspect signals 111
three outlet points 25, 98
thermal resistance 34
thermal switch 150
tie-bar 18
time delay fuse 33
timer (station stop) 96
toe of points 18

toggle switch 62, 68, 77
torque 45
track 12
track isolation 89
track joining 13
track length 13
track pickup 119
track separation 12
track spacing 12
transformer 29, 60, 151
triple pole, double throw (3 PDT) 79
triple slip 25
trunking 58
turntable 29, 98
turnout 17, 19,
two aspect signal 108
two wire signals 110
twisted cable 55

underhang 14
uncouplers 103

voltage regulator 41

watts 40
wave form 32, 37
web 21
welding simulator 117
wire 52, 80
wire clips 57

'Y' outlet points 25, 98

zebra crossing 118
zener diode 35, 147
zener voltage 35